*Khalid Y. Al-Qahtani and
Ali Elkamel*

**Planning and Integration of
Refinery and Petrochemical
Operations**

Related Titles

Reniers, G. L. L.

Multi-Plant Safety and Security Management in the Chemical and Process Industries

2010

ISBN: 978-3-527-32551-1

Lieberman, N.

Troubleshooting Process Plant Control

ISBN: 978-0-470-42514-5

Georgiadis, M., Kikkinides, E. S., Pistikopoulos, E. (eds.)

Process Systems Engineering

Volume 5: Energy Systems Engineering

2008

ISBN: 978-3-527-31694-6

Elvers, B. (ed.)

Handbook of Fuels

Energy Sources for Transportation

2008

ISBN: 978-3-527-30740-1

Papageorgiou, L., Georgiadis, M. (eds.)

Process Systems Engineering

Volume 3: Supply Chain Optimization

2008

ISBN: 978-3-527-31693-9

Papageorgiou, L., Georgiadis, M. (eds.)

Process Systems Engineering

Volume 4: Supply Chain Optimization

2008

ISBN: 978-3-527-31906-0

Wiley

Wiley Critical Content

Petroleum Technology, 2 Volume Set

ISBN: 978-0-470-13402-3

Bloch, H. P.

A Practical Guide to Compressor Technology

2006

ISBN: 978-0-471-72793-4

Ocic, O.

Oil Refineries in the 21st Century

Energy Efficient, Cost Effective, Environmentally Benign

2005

ISBN: 978-3-527-31194-1

Khalid Y. Al-Qahtani and Ali Elkamel

Planning and Integration of Refinery and Petrochemical Operations

WILEY-VCH Verlag GmbH & Co. KGaA

The Authors

Prof. Khalid Y. Al-Qahtani
Saudi Aramco
Process & Control Systems Dept
R-E-2790, Engin. Bldg (728A)
31311 Dhahran
Saudi Arabien

Prof. Ali Elkamel
University of Waterloo
Dept. of Chemical Engineering
University Avenue West 200
Waterloo, ON N2L 3G1
Kanada

All books published by Wiley-VCH are carefully produced. Nevertheless, authors, editors, and publisher do not warrant the information contained in these books, including this book, to be free of errors. Readers are advised to keep in mind that statements, data, illustrations, procedural details or other items may inadvertently be inaccurate.

Library of Congress Card No.: applied for

British Library Cataloguing-in-Publication Data
A catalogue record for this book is available from the British Library.

Bibliographic information published by the Deutsche Nationalbibliothek
The Deutsche Nationalbibliothek lists this publication in the Deutsche Nationalbibliografie; detailed bibliographic data are available on the Internet at http://dnb.d-nb.de.

© 2010 WILEY-VCH Verlag GmbH & Co. KGaA, Boschstr. 12, 69469 Weinheim, Germany

All rights reserved (including those of translation into other languages). No part of this book may be reproduced in any form – by photoprinting, microfilm, or any other means – nor transmitted or translated into a machine language without written permission from the publishers. Registered names, trademarks, etc. used in this book, even when not specifically marked as such, are not to be considered unprotected by law.

Cover Design Formgeber, Eppelheim
Typesetting Thomson Digital, Noida, India
Printing and Binding Strauss GmbH, Mörlenbach

Printed in the Federal Republic of Germany
Printed on acid-free paper

ISBN: 978-3-527-32694-5

Contents

Preface *IX*

Part One Background *1*

1 **Petroleum Refining and Petrochemical Industry Overview** *3*
1.1 Refinery Overview *3*
1.2 Mathematical Programming in Refining *5*
1.3 Refinery Configuration *7*
1.3.1 Distillation Processes *7*
1.3.2 Coking and Thermal Processes *8*
1.3.3 Catalytic Processes *9*
1.3.3.1 Cracking Processes *9*
1.3.3.2 Alteration Processes *9*
1.3.4 Treatment Processes *10*
1.3.5 Product Blending *10*
1.4 Petrochemical Industry Overview *11*
1.5 Petrochemical Feedstock *12*
1.5.1 Aromatics *12*
1.5.2 Olefins *13*
1.5.3 Normal Paraffins and Cyclo-Paraffins *13*
1.6 Refinery and Petrochemical Synergy Benefits *14*
1.6.1 Process Integration *14*
1.6.2 Utilities Integration *15*
1.6.3 Fuel Gas Upgrade *16*
References *16*

Part Two Deterministic Planning Models *19*

2 **Petroleum Refinery Planning** *21*
2.1 Production Planning and Scheduling *21*
2.2 Operations Practices in the Past *23*

Planning and Integration of Refinery and Petrochemical Operations. Khalid Y. Al-Qahtani and Ali Elkamel
Copyright © 2010 WILEY-VCH Verlag GmbH & Co. KGaA, Weinheim
ISBN: 978-3-527-32694-5

2.3	Types of Planning Models 24
2.4	Regression-Based Planning: Example of the Fluid Catalytic Cracker 24
2.4.1	Fluid Catalytic Cracking Process 25
2.4.2	Development of FCC Process Correlation 27
2.4.3	Model Evaluation 31
2.4.4	Integration within an LP for a Petroleum Refinery 31
2.5	Artificial-Neural-Network-Based Modeling: Example of Fluid Catalytic Cracker 36
2.5.1	Artificial Neural Networks 36
2.5.2	Development of FCC Neural Network Model 37
2.5.3	Comparison with Other Models 39
2.6	Yield Based Planning: Example of a Single Refinery 44
2.6.1	Model Formulation 46
2.6.1.1	Limitations on Plant Capacity 46
2.6.1.2	Material Balances 46
2.6.1.3	Raw Material Limitation and Market Requirement 47
2.6.1.4	Objective Function 47
2.6.2	Model Solution 48
2.6.3	Sensitivity Analysis 49
2.7	General Remarks 52
	References 53
3	**Multisite Refinery Network Integration and Coordination** 55
3.1	Introduction 55
3.2	Literature Review 57
3.3	Problem Statement 60
3.4	Model Formulation 61
3.4.1	Material Balance 62
3.4.2	Product Quality 63
3.4.3	Capacity Limitation and Expansion 64
3.4.4	Product Demand 65
3.4.5	Import Constraint 65
3.4.6	Objective Function 65
3.5	Illustrative Case Study 66
3.5.1	Single Refinery Planning 66
3.5.2	Multisite Refinery Planning 69
3.5.2.1	Scenario-1: Single Feedstock, Multiple Refineries with No Integration 70
3.5.2.2	Scenario-2: Single Feedstock, Multiple Refineries with Integration 71
3.5.2.3	Scenario-3: Multiple Feedstocks, Multiple Refineries with Integration 72
3.5.2.4	Scenario-4: Multiple Feedstocks, Multiple Refineries with Integration and Increased Market Demand 74

3.6	Conclusion 75	
	References 77	
4	**Petrochemical Network Planning** 81	
4.1	Introduction 81	
4.2	Literature Review 82	
4.3	Model Formulation 83	
4.4	Illustrative Case Study 84	
4.5	Conclusion 87	
	References 88	
5	**Multisite Refinery and Petrochemical Network Integration** 91	
5.1	Introduction 91	
5.2	Problem Statement 93	
5.3	Model Formulation 95	
5.4	Illustrative Case Study 99	
5.5	Conclusion 105	
	References 106	

Part Three Planning Under Uncertainty   109

6	**Planning Under Uncertainty for a Single Refinery Plant** 111	
6.1	Introduction 111	
6.2	Problem Definition 112	
6.3	Deterministic Model Formulation 112	
6.4	Stochastic Model Formulation 114	
6.4.1	Appraoch 1: Risk Model I 114	
6.4.1.1	Sampling Methodolgy 115	
6.4.1.2	Objective Function Evaluation 115	
6.4.1.3	Variance Calculation 116	
6.4.2	Approach 2: Expectation Model I and II 117	
6.4.2.1	Demand Uncertainty 117	
6.4.2.2	Process Yield Uncertainty 118	
6.4.3	Approach 3: Risk Model II 119	
6.4.4	Approach 4: Risk Model III 120	
6.5	Analysis Methodology 121	
6.5.1	Model and Solution Robustness 121	
6.5.2	Variation Coefficient 122	
6.6	Illustrative Case Study 123	
6.6.1	Approach 1: Risk Model I 124	
6.6.2	Approach 2: Expectation Models I and II 125	
6.6.3	Approach 3: Risk Model II 126	
6.6.4	Approach 4: Risk Model III 133	
6.7	General Remarks 133	
	References 137	

7	**Robust Planning of Multisite Refinery Network**	*139*
7.1	Introduction	*139*
7.2	Literature Review	*140*
7.3	Model Formulation	*142*
7.3.1	Stochastic Model	*142*
7.3.2	Robust Model	*144*
7.4	Sample Average Approximation (SAA)	*146*
7.4.1	SAA Method	*146*
7.4.2	SAA Procedure	*147*
7.5	Illustrative Case Study	*148*
7.5.1	Single Refinery Planning	*148*
7.5.2	Multisite Refinery Planning	*153*
7.6	Conclusion	*159*
	References	*159*
8	**Robust Planning for Petrochemical Networks**	*161*
8.1	Introduction	*161*
8.2	Model Formulation	*162*
8.2.1	Two-Stage Stochastic Model	*162*
8.2.2	Robust Optimization	*163*
8.3	Value to Information and Stochastic Solution	*165*
8.4	Illustrative Case Study	*166*
8.4.1	Solution of Stochastic Model	*167*
8.4.2	Solution of the Robust Model	*168*
8.5	Conclusion	*170*
	References	*171*
9	**Stochastic Multisite Refinery and Petrochemical Network Integration**	*173*
9.1	Introduction	*173*
9.2	Model Formulation	*174*
9.3	Scenario Generation	*177*
9.4	Illustrative Case Study	*177*
9.5	Conclusion	*181*
	References	*181*

Appendix A: Two-Stage Stochastic Programming *183*

Appendix B: Chance Constrained Programming *185*

Appendix C: SAA Optimal Solution Bounding *187*

Index *189*

Preface

Petroleum refining and the petrochemical industry account for a major share of the world energy and industrial market. In many situations, they represent the economic back-bone of industrial countries. Today, the volatile environment of the market and the continuous change in customer requirements lead to constant pressure to seek opportunities that properly align and coordinate the different components of the industry. In particular, petroleum refining and petrochemical industry coordination and integration is gaining a great deal of interest. Previous attempts in the field either studied the two systems in isolation or assumed limited interactions between them.

This book aims at providing the reader with a detailed understanding of the planning, integration and coordination of multisite refinery and petrochemical networks using proper deterministic and stochastic techniques. The book consists of three parts:

- **Part 1**: Background
- **Part 2**: Deterministic Planning Models
- **Part 3**: Planning under Uncertainty

Part 1, comprised of one chapter, introduces the reader to the configuration of petroleum refining and the petrochemical industry. It also discusses key classifications of petrochemical industry feedstock from petroleum products. The final part explains and proposes possible synergies between the petroleum refinery and the petrochemical industry.

Part 2, comprised of four chapters, focusses on the area of planning in petroleum refining and the petrochemical industry under deterministic conditions. Chapter 2 discusses the model classes used in process planning (i.e., empirical models, and first principle models) and provides a series of case studies to illustrate the concepts and impeding assumptions of the different modeling approaches. Chapter 3 tackles the integration and coordination of a multisite refinery network. It addresses the design and analysis of multisite integration and coordination strategies within a network of petroleum refineries through a mixed-integer linear programming (MILP) technique. Chapter 4 explains the general representation of a petrochemical planning model which selects the optimal network from the overall petrochemical superstructure. The system is modeled as a MILP problem and is illustrated via a numerical example. Chapter 5 addresses the integration between the multisite refinery system and the petrochemical industry. The chapter develops a framework for the design and analysis of possible integration and coordination strategies of multisite refinery and petrochemical networks to satisfy given petroleum and chemical product demand. The main feature of the proposed approach is the development of a methodology for the simultaneous analysis of process network integration within a multisite refinery and petrochemical system. Part 2 of this book serves as a foundation for the reader of Part 3.

Part 3, comprised of four chapters, tackles the area of planning in the petroleum refinery and the petrochemical industry under uncertainty. Chapter 6 explains the use of two-stage stochastic programming and the incorporation of risk management for a single site refinery plant. The example used in this chapter is simple enough for the reader to grasp the concept of two-stage stochastic programming and risk management and to be prepared for the larger scale systems in the remaining chapters. Chapter 7 extends the proposed model in Chapter 3 to account for model uncertainty by means of two-stage stochastic programming. Parameter uncertainty was considered and included coefficients of the objective function and right-hand-side parameters in the inequality constraints. Robustness is analyzed based on both model robustness and solution robustness, where each measure is assigned a scaling factor to analyze the sensitivity of the refinery plan and the integration network due to variations. The proposed technique makes use of the sample average approximation (SAA) method with statistical bounding techniques to give an insight on the sample size required to give adequate approximation of the problem. Chapter 8 addresses the planning, design and optimization of a network of petrochemical processes under uncertainty and robust considerations. Similar to the previous chapter, robustness is analyzed based on both model robustness and solution robustness. Parameter uncertainty considered in this part includes process yield, raw material and product prices, and lower product market demand. The expected value of perfect information (EVPI) and the value of the stochastic solution (VSS) are also investigated to illustrate numerically the value of including the randomness of the different model parameters. Chapter 9 extends the petroleum refinery and petrochemical industry integration problem, explained in Chapter 5, to consider different sources of uncertainties in model parameters. Parameter uncertainty considered includes imported crude oil price, refinery product price, petrochemical product price, refinery market demand, and petrochemical lower level product demand. The sample average approximation (SAA) method is within an iterative scheme to generate the required scenarios and provide solution quality by measuring the optimality gap of the final solution.

All chapters are equipped with clear figures and tables to help the reader understand the included topics. Furthermore, several appendices are included to explain the general background in the area of stochastic programming, chance constraint programming and robust optimization.

Part One
Background

1
Petroleum Refining and Petrochemical Industry Overview

Petroleum refining and the petrochemical industry account for a major share in the world energy and industrial market. In many situations, they represent the economic back-bone of industrial countries. Today, the volatile environment of the market and the continuous change in customer requirements lead to constant pressure to seek opportunities that properly align and coordinate the different components of the industry. In particular, petroleum refining and petrochemical industry coordination and integration is gaining a great deal of interest.

In this chapter, we will give an overview of the process configurations of petroleum refining and the petrochemical industry. We will also discuss the key classifications of petrochemical industry feedstock from petroleum products and explain and propose possible synergies between the petroleum refinery and the petrochemical industry.

1.1
Refinery Overview

The first refinery was built in Titusville, Pennsylvania in 1860 at a cost of $15 000 (Nelson, 1958). This refinery and other refineries at that time only used batch distillation to separate kerosene and heating oil from other crude fractions. During the early years, refining separation was performed using batch processing. However, with the increase in demand for petroleum products, continuous refining became a necessity. The first widely recognized continuous refinery plants emerged around 1912 (Nelson, 1958). With the diversity and complexity of the demand for petroleum products, the refining industry has developed from a few simple processing units to very complex production systems. A simplified process flow diagram of a typical modern refinery is shown in Figure 1.1. For a detailed history of the evolution of refining technologies, we refer the reader to Nelson (1958) and Wilson (1997).

Typically, a refinery is made up of several distinct components that constitute a total production system, as shown in Figure 1.2. These components include:

4 | *1 Petroleum Refining and Petrochemical Industry Overview*

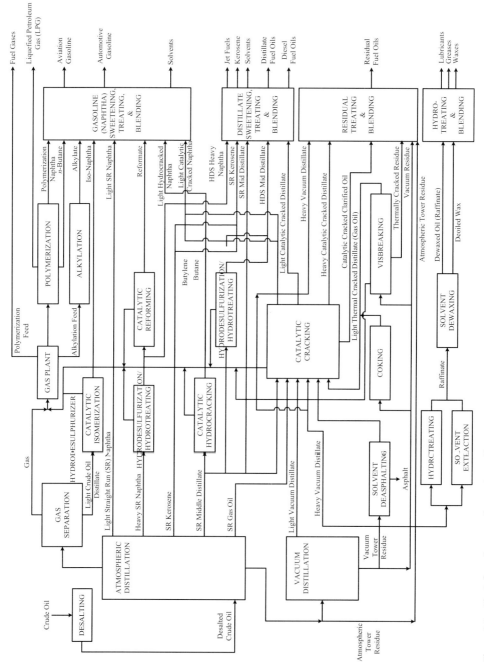

Figure 1.1 Block flow diagram of a modern refinery.

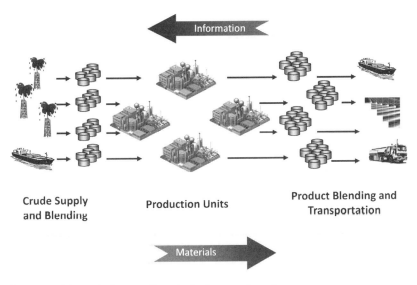

Figure 1.2 Schematic diagram of standard refinery configuration.

- **Crude Supply and Blending:** This area includes receiving facilities and a tank area (tank farm) where all crude oil types are received and either blended or sent directly to the production system.
- **Production Units:** Production units separate crude oil into different fractions or cuts, upgrade and purify some of these cuts, and convert heavy fractions to light, more useful fractions. This area also includes the utilities which provide the refinery with fuel, flaring capability, electricity, steam, cooling water, fire water, sweet water, compressed air, nitrogen, and so on, all of which are necessary for the safe operation of the refinery.
- **Product Blending and Transportation:** In this area the final products are processed according to either predetermined recipes and/or to certain product specifications. This area also includes the dispatch (terminals) of finished products to the different customers.

1.2
Mathematical Programming in Refining

The petroleum industry has long made use of mathematical programming and its different applications. The invention of both the simplex algorithm by Dantzig in 1947 and digital computers was the main driver for the widespread use of linear programming (LP) applications in the industry (Bodington and Baker, 1990). Since then, many early applications followed in the area of refinery planning (Symonds, 1955; Manne, 1958; Charnes and Cooper, 1961; Wagner, 1969; Addams and Griffin, 1972) and distribution planning (Zierer, Mitchell and White, 1976).

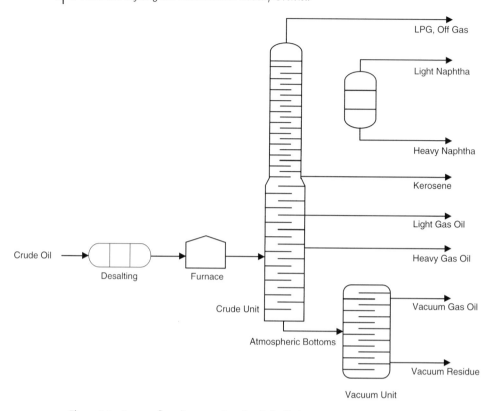

Figure 1.3 Process flow diagram of crude oil distillation process.

One of the main challenges that inspired more research in the area of refining was the blending or pooling problem (Bodington and Baker, 1990). The inaccurate and inconsistent results from the use of linear blending relations led to the development of many techniques to handle nonlinearities. The nonlinearities arise mainly because product properties, such as octane number and vapor pressure, assume a nonlinear relationship of quantities and properties of each blending component (Lasdon and Waren, 1983). In this context, we will describe two commonly used approaches in industry and commercial planning softwares to tackle this problem. They are linear blending indices and successive linear programming (SLP).

Linear blending indices are dimensionless numerical figures that were developed to represent true physical properties of mixtures on either a volume or weight average basis (Bodington and Baker, 1990). They can be used directly in the LP model and span the most important properties in petroleum products, including octane number, pour point, freezing point, viscosity, sulfur content, and vapor pressure. Many refineries and researchers use this approximation. Blending indices tables and graphs can often be found in petroleum refining books such as Gary and Handwerk (1994) or can be proprietorially developed by refining companies for their own use.

Successive linear programming, on the other hand, is a more sophisticated method to linearize blending nonlinearities in the pooling problem. The idea of SLP was first introduced by Griffith and Stewart (1961) of the Shell Oil Company where it was named the method of approximation programming (MAP). They utilized the idea of a Taylor series expansion to remove nonlinearities in the objective function and constraints then solving the resulting linear model repeatedly. Every LP solution is used as an initial solution point for the next model iteration until a satisfying criterion is reached. Bounding constraints were added to ensure the new model feasibility. Following their work, many improvement heuristics and solution algorithms were developed to accommodate bigger and more complex problems (Lasdon and Waren, 1980). Most commercial blending softwares and computational tools nowadays are based on SLP, such as RPMS by Honeywell Process Solutions (previously Booner and Moore, 1979) and PIMS by Aspen Technology (previously Bechtel Corp., 1993). However, such commercial tools are not built to support studies on capacity expansion alternatives, design of plants integration and stochastic modeling and analysis.

All in all, the petroleum industry has invested considerable effort in developing sophisticated mathematical programming models to help planners provide overall planning schemes for refinery operations, crude oil evaluation, and other related tasks.

1.3
Refinery Configuration

1.3.1
Distillation Processes

Crude oil distillation is the heart of and major unit in the refinery. Distillation is used to separate oil into fractions by distillation according to their boiling points. Prior to distillation, crude oil is first treated to remove salt content, if higher than 10 lb/1000 bbl, using single or multiple desalting units. This is required in order to minimize corrosion and fouling in the downstream heating trains and distillation columns. As illustrated in Figure 1.3, distillation is usually divided into two steps, atmospheric and vacuum fractionation according to the pressure at which fractionation is achieved. This is done in order to achieve higher separation efficiencies at a lower cost. After heating the crude to near its boiling point, it is introduced to the distillation column in which vapor rising through trays in the column is in direct contact with down-flowing liquid on the trays. During this process, higher boiling point fractions in the vapor phase are condensed and lighter fractions in the liquid are vaporized. This continuous process allows the various fractions of the crude oil with similar boiling points to achieve equilibrium and separate. Liquid can then be drawn off the column at different heights as product and sent for further treating or storage. Common products from the atmospheric distillation column include liquefied petroleum gas (LPG), naphtha, kerosene, gas oils and heavy residues.

The atmospheric bottom, also known as reduced oil, is then sent to the vacuum unit where it is further separated into vacuum gas oil and vacuum residues. Vacuum distillation improves the separation of gas oil distillates from the reduced oil at temperatures less than those at which thermal cracking would normally take place. The basic idea on which vacuum distillation operates is that, at low pressure, the boiling points of any material are reduced, allowing various hydrocarbon components in the reduced crude oil to vaporize or boil at a lower temperature. Vacuum distillation of the heavier product avoids thermal cracking and hence product loss and equipment fouling.

1.3.2
Coking and Thermal Processes

Nowadays more refineries are seeking lighter and higher quality products out of the heavy residues. Coking and other thermal processes convert heavy feedstocks, usually from distillation processes, to more desirable and valuable products that are suitable feeds for other refinery units. Such units include coking and visbreaking.

One of the widely used coking processes is delayed coking. It involves severe thermal cracking of heavy residues such as vacuum oil, thermal tars, and sand bitumen. The actual coking in this process takes place in the heater effluent surge drum and for this reason the process is called "delayed coking". The coke produced by this process is usually a hard and porous sponge-like material. This type of coke is called sponge coke and exists in a range of sizes and shapes. Many other types of coke are commercially available in the market and have a wide range of uses, see Table 1.1. Other coking processes, including flexicoking and fluid coking, have been developed by Exxon.

The other thermal cracking process is visbreaking. This is a milder thermal process and is mainly used to reduce the viscosities and pour points of vacuum residues to

Table 1.1 End use of coke products (Gary and Handwerk, 2001).

Application	Coke type	End use
Carbon source	Needle	Electrodes
		Synthetic graphite
	Sponge	Aluminum anodes
		TiO2 pigments
		Carbon raiser
		Silicon carbide
		Foundries
		Coke ovens
Fuel use	Sponge	Space heating in Europe/Japan
		Industrial boilers
	Shot	Utilities
	Fluid	Cogeneration
	Flexicoke	Lime
		Cement

meet some types of fuel oil specifications and also to increase catalytic cracker feedstock. The two widely used processes in visbreaking are coil visbreaking and soaker visbreaking. In coil visbreaking most of the cracking takes place in the furnace coil whereas in soaker visbreaking, cracking takes place in a drum downstream of the heater, called the soaker. Each process offers different advantages depending on the given situation.

1.3.3
Catalytic Processes

There are two types of catalytic conversion units in the refinery, cracking and alteration processes. Catalytic cracking converts heavy oils into lighter products that can be blended to produce high value final products, such as gasoline, jet fuels and diesel. Whereas, catalytic altering processes convert feedstocks to higher quality streams by rearranging their structures. These processes include reforming, alkylation and isomerization units. Catalytic processes produce hydrocarbon molecules with double bonds and form the basis of the petrochemical industry.

1.3.3.1 Cracking Processes
Cracking processes mainly include catalytic cracking and hydrocracking. Catalytic cracking involves breaking down and rearranging complex hydrocarbons into lighter molecules in order to increase the quality and quantity of desirable products such as kerosene, gasoline, LPG, heating oil, and petrochemical feedstock. Catalytic cracking follows a similar concept to thermal cracking except that catalysts are used to promote and control the conversion of the heavier molecules into lighter products under much less severe operating conditions. The most commonly used process in the industry is fluid catalytic cracking (FCC) in which oil is cracked in a fluidized catalyst bed where it is continuously circulated between the reaction state and the regeneration state.

Hydrocracking on the other hand is a process that combines catalytic cracking and hydrogenation where the feed is cracked in the presence of hydrogen to produce more desirable products. This process mainly depends on the feedstock characteristics and the relative rates of the two competing reactions, hydrogenation and cracking. In the case where the feedstock has more paraffinic content, hydrogen acts to prevent the formation of polycyclic aromatic compounds. Another important role of hydrogen is to reduce tar formation and prevent buildup of coke on the catalyst.

1.3.3.2 Alteration Processes
Alteration processes involve rearranging feed stream molecular structure in order to produce higher quality products. One of the main processes in this category is catalytic reforming. Reforming is an important process used to convert low-octane feedstock into high-octane gasoline blending components called reformate. The kinetics of reforming involves a wide range of reactions such as cracking, polymerization, dehydrogenation, and isomerization taking place simultaneously. Depending on the properties of the feedstock, measured by the paraffin, olefin, naphthene,

and aromatic content (PONA), and catalysts used, reformates can be produced with very high concentrations of toluene, benzene, xylene, and other aromatics. Hydrogen, a by-product of the reforming process, is separated from the products and used as a feed in other refining processes.

Another alteration process is alkylation which is used to produce higher octane aviation gasoline and petrochemical feedstock for explosives and synthetic rubber. An isomerization process is also used to produce more material as an alkylation feedstock.

1.3.4
Treatment Processes

Treatment processes are applied to remove impurities, and other constituents that affect the properties of finished products or reduce the efficiency of the conversion processes. A typical example of a treating process is hydrotreating.

Catalytic hydrotreating is a hydrogenation process used to remove about 90% of contaminants such as nitrogen, sulfur, oxygen, and metals from liquid petroleum fractions. These contaminants, if not removed from the petroleum fractions, can have a negative impact on the equipment, the catalysts, and the quality of the finished product. Hydrotreating is mainly used prior to catalytic reforming to reduce catalyst contamination and before catalytic cracking to reduce sulfur and improve product yields. It is also used to upgrade middle-distillate petroleum fractions into finished kerosene, diesel fuel, and heating fuel oils and converts olefins and aromatics to saturated compounds. One of the emerging applications of this process is the treatment of pyrolysis gasoline (pygas), a by-product from the manufacture of ethylene, in order to improve its quality. Typically, the outlet for pygas has been motor gasoline blending, a suitable route in view of its high octane number. However, only small portions can be blended untreated owing to the unacceptable odor, color, and gum-forming tendencies of this material. The quality of pygas, which is high in diolefin content, can be satisfactorily improved by hydrotreating, whereby conversion of diolefins into mono-olefins provides an acceptable product for motor gas blending.

1.3.5
Product Blending

Blending is the process of mixing hydrocarbon fractions, additives, and other components to produce finished products with specific properties and desired characteristics. Products can be blended in-line through a manifold system, or batch blended in tanks and vessels. In-line blending of gasoline, distillates, jet fuel, and kerosene is accomplished by injecting proportionate amounts of each component into the main stream where turbulence promotes thorough mixing. Additives, including octane enhancers, metal deactivators, anti-oxidants, anti-knock agents, gum and rust inhibitors, detergents, and so on, are added during and/or after blending to provide specific properties not inherent in hydrocarbons.

1.4
Petrochemical Industry Overview

The petrochemical industry is a network of highly integrated production processes. The products of one plant may have an end use but may also represent raw materials for another process. Most chemicals can be produced by many different sequences of reactions and production processes. This multiplicity of production schemes offers the opportunity for switching between production methods and raw materials utilization.

Petroleum feedstock, natural gas and tar represent the main production chain drivers for the petrochemical industry (Bell, 1990). From these, many important petrochemical intermediates are produced, including ethylene, propylene, butylenes, butadiene, benzene, toluene, and xylene. These essential intermediates are then converted to many other intermediates and final petrochemical products, constructing a complex petrochemical network. Figure 1.4 depicts a portion of the petrochemical alternative routes to produce cellulous acetate.

Figure 1.4 is in fact a small extract from much larger and comprehensive flow diagrams found in Stanford Research Institute (SRI) reports. Note that certain chemicals, acetaldehyde and acetic acid for example, appear in more than one place in the flowchart. This reflects the different alternative production routes available for most chemicals. In the industry, many chemicals are products of more than one

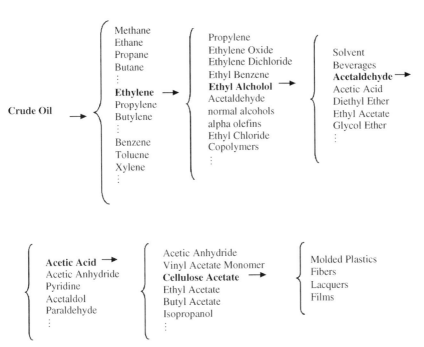

Figure 1.4 A Single Route of petroleum feedstock to petrochemical products (Bell, 1990).

process, depending upon local conditions, corporate polices, and desired by-products (Bell, 1990).

The flexibility in the petrochemical industry production and the availability of many process technologies require adequate strategic planning and a comprehensive analysis of all possible production alternatives. Therefore, a model is needed to provide the development plan of the petrochemical industry. The model should account for market demand variability, raw material and product price fluctuations, process yield inconsistencies, and adequate incorporation of robustness measures.

The realization of the need and importance of petrochemical planning has inspired a great deal of research in order to devise different models to account for the overall system optimization. Optimization models include continuous and mixed-integer programming under deterministic or parameter uncertainty considerations. Related literature is reviewed at a later stage in this book, based on the chapter topic.

1.5
Petrochemical Feedstock

The preparation of intermediate petrochemical streams requires different processing alternatives depending on the feedstock quality. In our classification of petrochemical feedstocks we closely follow that of Gary and Handwerk (1994) consisting of aromatics, olefins, and paraffin/cyclo-paraffin compounds. The classification of petrochemical feedstocks into these clusters helps to identify the different sources in the refinery that provide suitable feedstock and, therefore, to better recognize areas of synergy between the refinery and petrochemical systems.

1.5.1
Aromatics

Aromatics are hydrocarbons containing a benzene ring which is a stable and saturated compound. Aromatics used by the petrochemical industry are mainly benzene, toluene, xylene (BTX) as well as ethylbenzene and are produced by catalytic reforming where their yield increases with the increase in reforming severity (Gary and Handwerk, 1994). Extractive distillation by different solvents, depending on the chosen technology, is used to recover such compounds. BTX recovery consists of an extraction using solvents that enhances the relative volatilities of the preferred compound followed by a separation process based on the boiling points of the products. Further processing of xylenes using isomerization/separation processes is commonly required to produce o-, m-, and p-xylene mixtures, depending on market requirements. Benzene, in particular, is a source of a wide variety of chemical products. It is often converted to ethylbenzene, cumene, cyclohexane, and nitrobenzene, which in turn are further processed to other chemicals including styrene, phenol, and aniline (Rudd et al., 1981). Toluene production, on the other hand, is mainly driven by benzene and mixed xylenes demand. Mixed xylene, particularly in Asia, is used to produce *para*-xylene and polyester (Balaraman, 2006).

The other source of aromatics is the pyrolysis gasoline (pygas) which is a byproduct of naphtha or gas oil steam cracking. This presents an excellent synergistic opportunity between refinery, BTX complex and stream cracking for olefins production.

1.5.2
Olefins

Olefins are hydrocarbon compounds with at least two carbon atoms and having a double bond. Their unstable nature and tendency to polymerize makes them one of the very important building blocks for the chemical and petrochemical industry (Gary and Handwerk, 1994). Although olefins are produced by fluid catalytic cracking in refineries, the main production source is through steam cracking of liquefied petroleum gas (LPG), naphtha or gas oils.

The selection of steam cracker feedstock is mainly driven by market demand as different feedstock qualities produce different olefins yields. One of the commonly used feed quality assessment methods in practice is the Bureau of Mines Correlation Index (BMCI) (Gonzalo et al., 2004). This index is a function of average boiling point and specific gravity of a particular feedstock. The steam cracker feed quality improves with a decrease in the BMCI value. For instance, vacuum gas oil (VGO) has a high value of BMCI and, therefore, is not an attractive steam cracker feed. The commonly used feedstocks in industry are naphtha and gas oil.

Steam cracking plays an instrumental role in the petrochemical industry in terms of providing the main petrochemical intermediates for the downstream industry. The steam cracker olefin production includes ethylene, propylene, butylene and benzene. These intermediates are further processed into a wide range of polymers (plastics), solvents, fibers, detergents, ammonia and other synthetic organic compounds for general use in the chemical industry (Rudd et al., 1981). In a situation where worldwide demand for these basic olefins is soaring, more studies are being conducted to maximize steam cracking efficiency (Ren, Patel and Blok, 2006). An alternative strategy would be to seek integration possibilities with the refinery as they both share feedstocks and products that can be utilized to maximize profit and processing efficiency.

1.5.3
Normal Paraffins and Cyclo-Paraffins

Paraffin hydrocarbon compounds contain only single bonded carbon atoms which give them higher stability. Normal paraffin compounds are abundantly present in petroleum fractions but are mostly recovered from light straight-run (LSR) naphtha and kerosene. However, the non-normal hydrocarbon components of LSR naphtha have a higher octane number and, therefore, are preferred for gasoline blending (Meyers, 1997). For this reason, new technologies have been developed to further separate LSR naphtha into higher octane products that can be used in the gasoline pool and normal paraffins that are used as steam cracker feedstock (e.g., UOP IsoSiv™ Process). The normal paraffins recovered from kerosene, on the other hand, are mostly used in biodegradable detergent manufacturing.

Cyclo-paraffins, also referred to as naphthenes, are mainly produced by dehydrogenation of their equivalent aromatic compounds; such as the production of cyclohexane by dehydrogenation of benzene. Cyclohexane is mostly used for the production of adipic acid and nylon manufacturing (Rudd et al., 1981).

1.6
Refinery and Petrochemical Synergy Benefits

Process integration in the refining and petrochemical industry includes many intuitively recognized benefits of processing higher quality feedstocks, improving the value of byproducts, and achieving better efficiencies through sharing of resources. Table 1.2 illustrates different refinery streams that can be of superior quality when used in the petrochemical industry. The potential integration alternatives for refining and petrochemical industries can be classified into three main categories; (i) process integration, (ii) utilities integration, and (iii) fuel gas upgrade. The integration opportunities discussed below are for a general refinery and a petrochemical complex. Further details and analysis of the system requirements can be developed based on the actual system infrastructure, market demand, and product and energy prices.

1.6.1
Process Integration

The innovative design of different refinery processes while considering the downstream petrochemical industry is an illustration of the realization of refining and

Table 1.2 Petrochemical alternative use of refinery streams (Anon, 1998).

Refinery Stream	Petrochemical Stream	Alternative Refinery Use
FCC offgas	Ethylene	Fuel gas
Refinery propylene (FCC)	Propylene	Alkylation/polygasoline
Reformate	Benzene, toluene, xylenes	Gasoline blending
Naphtha and LPG	Ethylene	Gasoline Blending
Dilute ethylene (FCC and delayed coker offgas)	Ethylbenzene	Fuel gas
Refinery propylene (FCC product)	Polypropylene, Cumene, Isopropanol, Oligomers	Alkylation
Butylenes (FCC and delayed coker)	MEK (methyl ethyl ketone)	Alkylation, MTBE
Butylenes (FCC and delayed coker)	MTBE	Alkylation, MTBE
Refinery benzene and hydrogen	Cyclohexane	Gasoline blending
Reformate	o-xylene	Gasoline blending
Reformate	p-xylene	Gasoline blending
Kerosine	n-paraffins	Refinery product
FCC light cycle oil	Naphthalene	Diesel blending

petrochemical integration benefits. This is demonstrated by the wide varieties of refinery cracking and reforming technologies that maximize olefin production. Some of the available technologies include cracking for high propylene and gasoline production (Fujiyama et al., 2005), maximum gasoline and LPG production, and low-pressure combination-bed catalytic reforming for aromatics (Wang, 2006). Other technologies include different extractive treatments of refinery streams, for example, aromatic recovery from light straight-run (LSR) naphtha. The normal paraffins of the LSR, on the other hand, are typically used as a steam cracker feedstock (Meyers, 1997).

Reforming, as mentioned above, is the main source of aromatics in petroleum refining where their yield increases with the increase in reforming severity. Aromatics in the reformate streams are recovered by extractive distillation using different solvents, depending on the chosen technology. The benzene-toluene-xylenes (BTX) complex is one of the petrochemical processes that makes use of many of the benefits of integration with petroleum refining. The integration benefits are not only limited to the process side but also extend to the utilities, as will be explained in the following section.

Pyrolysis gasoline (pygas), a byproduct of stream cracking, can be further processed in the BTX complex to recover the aromatic compounds and the raffinate after extraction can be blended in the gasoline or naphtha pool (Balaraman, 2006). If there is no existing aromatics complex to further process the pygas, it could alternatively be routed to the reformer feed for further processing (Philpot, 2007). However, this alternative may not be viable in general as most reformers run on maximum capacity. Pygas from steam cracking contains large amounts of diolefins which are undesirable due to their instability and tendency to polymerize yielding filter plugging compounds. For this reason, hydrogenation of pygas is usually recommended prior to further processing.

1.6.2
Utilities Integration

Petroleum refining and the basic petrochemical industry are the most energy intensive processes in the chemical process industry (Ren, Patel and Blok, 2006). The energy sources in these processes assume different forms including fuel oil, fuel gas, electrical power, and both high and low pressure steam. The different energy requirements and waste from the whole range of refinery and petrochemical units present intriguing opportunities for an integrated complex. Integration of energy sources and sinks of steam cracking, for instance, with other industrial processes, particularly natural gas processing, can yield significant energy savings reaching up to 60% (Ren, Patel and Blok, 2006). Furthermore, gas turbine integration (GTI) between petrochemical units and ammonia plants can lead to a reduction in energy consumption by up to 10% through exhaust-heat recovery (Swaty, 2002). This can be readily extended to the refinery processing units which span a wide variety of distillation, cracking, reforming, and isomerization processes.

Hydrogen is another crucial utility that is receiving more attention recently, mainly due to the stricter environmental regulations on sulfur emissions. Reduction of sulfur emissions is typically achieved by deeper desulfurization of petroleum fuels which in turn requires additional hydrogen production (Crawford, Bharvani and Chapel, 2002). A less capital intensive alternative to alleviate hydrogen shortage is to operate the catalytic reformer at higher severity. However, higher severity reforming increases the production of BTX aromatics which consequently affect the gasoline pool aromatics specification. Therefore, the BTX extraction process becomes a more viable alternative for the sake of aromatics recovery as well as maintaining the gasoline pool within specification (Crawford, Bharvani and Chapel, 2002). The capital cost for the implementation of such a project would generally be lower as the BTX complex and refinery would share both process and utilities streams.

1.6.3
Fuel Gas Upgrade

Refinery fuel gas is generated from refinery processes and is mainly comprised of C_1/C_2 fractions and some hydrogen. Considerable amounts of light hydrocarbons are produced from the different conversion units in the refinery and are collected in the common fuel gas system. For instance, FCC off gas contains significant amounts of ethylene and propylene which can be extracted and processed as petrochemical feedstocks. A number of integrated U.S. and European refineries have recognized and capitalized on this opportunity by recovering these high value components (Swaty, 2002). This type of synergy requires proper planning and optimization between the petroleum refining and petrochemical complexes.

The other major component is hydrogen which typically accounts for 50–80% of the refinery fuel gas. This substantial amount of hydrogen is passed to the fuel gas system from different sources in the refinery. The most significant source, however, is the catalytic reforming. Hydrogen recovery using economically attractive technologies is of great benefit to both refineries and petrochemical systems, especially with the increasing strict environmental regulations on fuels.

References

Addams, F.G. and Griffin, J.M. (1972) Economic-linear programming model of the U.S. petroleum refining industry. *Journal of American Statistics Association*, 67, 542.

Anon (1998) Petrochemical complex shields refining profits, *Oil and Gas Journal*, 96, 31.

Balaraman, K.S. (2006) Maximizing hydrocarbon value chain by innovative concepts. *Journal of the Petrotech Society*, 3, 26.

Bechtel Corp. (1993) PIMS (Process Industry Modeling System). User's manual, version 6.0. Houston, TX.

Bell, J.T. (1990) Modeling of the Global Petrochemical Industry. Ph.D. Thesis. University of Wisconsin-Madison.

Bodington, C.E. and Baker, T.E. (1990) A history of mathematical programming in the petroleum industry. *Interfaces*, 20, 117.

Booner & Moore Management Science (1979) RPMS (Refinery and Petrochemical Modeling System): A system description. Houston, TX.

Charnes, A. and Cooper, W. (1961) *Management Models and Industrial Applications of Linear Programming*, John Wiley and Sons, New York.

Crawford, C.D., Bharvani, R.R., and Chapel, D.G. (2002) Integrating petrochemicals into the refinery. *Hydrocarbon Engineering*, **7**, 35.

Fujiyama, Y., Redhwi, H.H., Aitani, A.M., Saeed, M.R., and Dean, C.F. (2005) Demonstration plant for new FCC technology yields increased propylene. *Oil & Gas Journal*, **103**, 54.

Gary, J.H. and Handwerk, G.E. (1994) *Petroleum Refining: Technology and Economics*, Marcel Dekker Inc., New York.

Gonzalo, M.F., Balseyro, I.G., Bonnardot, J., Morel, F., and Sarrazin, P. (2004) Consider integrating refining and petrochemical operations. *Hydrocarbon Processing*, **83**, 61.

Griffith, R.E. and Stewart, R.A. (1961) A nonlinear programming technique for the optimization of continuous processes. *Management Science*, **7**, 379.

Lasdon, L.S. and Waren, A.D. (1980) Survey of nonlinear programming applications. *Operations Research*, **28**, 1029.

Lasdon, L.S. and Waren, A.D. (1983) Large scale nonlinear programming. *Computers & Chemical Engineering*, **7**, 595.

Manne, A. (1958) A linear programming model of the US petroleum refining industry. *Econometrica*, **26**, 67.

Meyers, R.A. (1997) *Handbook of Petroleum Refining Processes*, 2nd edn, The McGraw-Hill Companies Inc., USA.

Nelson, W.L. (1958) *Petroleum Refinery Engineering*, 4th edn, McGraw-Hill Book Company, New York.

Philpot, J. (2007) Integration profitability. *Hydrocarbon Engineering*, **12**, 41.

Ren, T., Patel, M. and Blok, K. (2006) Olefins from conventional and heavy feedstocks: Energy use in steam cracking and alternative processes. *Energy*, **31**, 425.

Rudd, D.F., Fathi-Afshar, S., Trevino, A.A., and Statherr, M.A. (1981) *Petrochemical Technology Assessment*, John Wiley & Sons, New York.

Swaty, T.E. (2002) Consider over-the-fence product stream swapping to raise profitability. *Hydrocarbon Processing*, **81**, 37.

Symonds, G.H. (1955) *Linear Programming-The Solution of Refinery Problems*, Esso Standard Oil Company, New York.

Wagner, H. (1969) *Principles of Operations Research with Applications to Managerial Decision*, Prentice-Hall Inc., New Jersey.

Wang, J. (2006) China's petrochemical technologies. *Chemistry International*, **28**, 1–4.

Wilson, J.W. (1997) *Fluid Catalytic Cracking-Technology and Operation*, PennWell Publishing Company, Tulsa, OK.

Zierer, T.K., Mitchell, W.A., and White, T.R. (1976) Practical applications of linear programming to Shell's distribution problems. *Interfaces*, **6**, 13.

Part Two
Deterministic Planning Models

2
Petroleum Refinery Planning

The petroleum refining industry accounts for a major share in the world energy and industrial market. Through proper planning and the use of adequate mathematical models for the different processing units, many profit improving opportunities can be realized and acquired. The development and selection of suitable mathematical modeling techniques is an art that requires practice, good understanding of the process interactions, and experience.

In this chapter, we give an overview of planning and scheduling practices in the refinery industry and discuss mathematical modeling classes used in process planning. We also provide a series of examples to illustrate the use of mathematical models in the planning of refinery processes. The readers should be able to apply and extend the same techniques explained in this chapter to their own refinery system. The chapter concludes with some reflections of the current difficulties and challenges in the planning of petroleum refineries.

2.1
Production Planning and Scheduling

Planning and scheduling can be defined as developing strategies for the allocation of equipment, utility or labor resources over time to execute specific tasks in order to produce single or several products (Grossmann, van den Heever and Harjunkoski, 2001). In most research dealing with planning and scheduling, there seems to be no clear cut division between the two. Hartmann (1998) and Grossmann, van den Heever and Harjunkoski (2001) pointed out some of the differences between a planning model and a scheduling model. In a general sense, process manufacturing planning models consider high level decisions such as investment on longer time horizons. This coarse aggregation approach results in a loss of manufacturing detail such as the sequence or the order in which specific manufacturing steps are executed. Scheduling models, on the other hand, are concerned more with the feasibility of the operations to accomplish a given number and order of tasks. Scheduling involves determining a feasible sequence and timing of production operations at different pieces of equipment so as to meet the production goals laid out by the planning

Planning and Integration of Refinery and Petrochemical Operations. Khalid Y. Al-Qahtani and Ali Elkamel
Copyright © 2010 WILEY-VCH Verlag GmbH & Co. KGaA, Weinheim
ISBN: 978-3-527-32694-5

model. A key characteristic of scheduling operations is the dynamic and extensive information required to describe the system. Customer orders, resource availability, and manufacturing processes undergo relatively rapid changes, resulting in a compelling need for efficient management of information and resources.

Planning problems can mainly be distinguished as strategic, tactical or operational, based on the decisions involved and the time horizon considered (Grossmann, van den Heever and Harjunkoski, 2001). The strategic level planning considers a time span of more than one year and covers the whole width of an organization. At this level, approximate and/or aggregate models are adequate and mainly consider future investment decisions. Tactical level planning typically involves a midterm horizon of a few months to a year, where the decisions usually include production, inventory, and distribution. Operational level covers shorter periods of time, spanning from one week to three months, where the decisions involve actual production and allocation of resources. For a general process operations hierarchy, planning is at the highest level of command. As shown in Figure 2.1, enterprise-wide planning provides production targets for each individual site, where each site transforms the plans into schedules and operational/control targets.

Despite the differences, it is obvious that production planning, scheduling, and operations control are all closely-related activities. Decisions made at the production planning level have a great impact at the scheduling level, while the scheduling in itself determines the feasibility of executing the production plans with the resulting

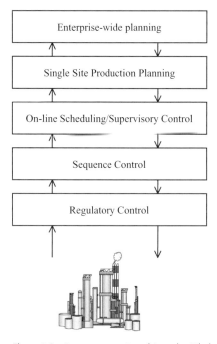

Figure 2.1 Process operations hierarchy (Shah, 1998).

decisions dictating operations control. Ideally, all three activities should be analyzed and optimized simultaneously, hence the need for integration between planning, scheduling, and operational activities, with the expectation that this would greatly enhance the overall performance of not only the refinery or process plant concerned, but also the parent governing organization. It is desirable to extend the scope of this hierarchy to include the highest level of strategic decision making, that is, the planning and design of production capacities required for future operations. While these levels can be viewed to constitute a hierarchy, the requirements of the hierarchy dictate that these levels communicate in a two-way interactive dynamics, with the lower levels communicating suitably aggregated performance limits and capacities to the upper levels. This is essentially the challenge of integrating the planning, scheduling, and operations functions of a process plant, primarily the flow of information between the various levels. In petroleum refineries, this problem stands out, for example, when dealing with activities such as crude oil procurement, logistics of transportation, and scheduling of processes. For more information on the development of such hierarchal operations and their challenges, we refer the interested readers to Reklaitis (1991), Rippin (1993), Shah (1998), and Grossmann (2005).

2.2
Operations Practices in the Past

In most non-integrated situations, strategic planning is performed by one entity close to the marketing and supply functions, but not part of them. Planning activities serve to consolidate feedstock purchases, commitments, and sales opportunities by attempting to set achievable targets for the plant. Scheduling is undertaken by another entity that stands between planning and operations. It attempts to produce a schedule that is feasible, if not optimal, to meet commitments. Process operations are handled by yet another entity, usually compartmentalized by processes, that operates the processes to the best capability, given the information available from planning and scheduling activities. The three entities have different objectives and possibly have different reward motivations and reward structures, which lead to different philosophies of what constitutes a job well done (Bodington, 1995).

The petroleum industry has invested considerable effort in developing sophisticated mathematical programming models to help planners provide strategies and directions for refinery operations, crude oil evaluation, and other related tasks. Likewise, there has been substantial development and implementation of tools for scheduling, as well as considerable efforts towards advanced process control for process plants to achieve optimal operations. Unfortunately, a gap will always exist between the three activities when working in isolation from each other. Typically, the refinery scheduler attempts to use the monthly linear programming (LP) model plan to develop a detailed day-to-day schedule based on scheduled crude and feedstock arrivals, product lifting, and process plant availabilities and constraints. The schedule usually includes details of the operation of each process unit, the transfer of intermediates to and from the tank farm, and product blending schedules. However, the scheduling is performed for each tank instead of for a pool. Moreover, most refinery schedulers have few extensive computing tools to accomplish this task. Many use spreadsheets that contain individual operating modes for the primary processes and for the main feedstocks, based on the same data employed in the LP model. The scheduler utilizes the spreadsheet to generate manufacturing plans on a daily or weekly basis. Compounding the problem is the fact that deficiencies in planning or operations often create problems that appear in the scheduling process. Operating deficiencies or inferior data on the status of the production process could potentially lead to customer service problems. These problems may also occur due to either a planning activity with an overly optimistic estimate of available capacity or a poor understanding of the production capabilities.

2.3
Types of Planning Models

Mathematical models can be classified in different ways depending on the type of analysis involved. For example, one can classify them according to their composing variables into linear and nonlinear models, or according to the nature of their variables and parameters into deterministic and stochastic models, or according to their component state into static and dynamic models. However, from a process engineering perspective, we are interested in the inherent representation of the mathematical model of the actual physical system. Based on this, there are two general model classes: (i) mechanistic models, and (ii) empirical models. Mechanistic models are those based on a theoretical understanding of the system and the interactions between its process variables. They are often based on the application of conservation principles (i.e., material and energy balances) and equilibrium relationships. The main advantage of such fundamental models is the ability to construct them prior to putting the system into operation. Empirical models, on the other hand, also known as black box and data driven models, are useful when mechanistic models are difficult to implement due to complexity or resource limitation. In empirical models the system is viewed in terms of its inputs, outputs, and the relation between them, without any knowledge of the internal mechanism of the system.

In the next sections, we will present a series of empirical model based case studies that illustrate different planning techniques commonly used in practice by many refineries.

2.4
Regression-Based Planning: Example of the Fluid Catalytic Cracker

In large-scale plants, such as refinery processing plants, even a small increase in yield can lead to a significant impact on profitability. This, and the increasing feed heaviness, as well as the more stringent demands on product qualities and environmental regulations (Anabtawi et al., 1996) make it necessary to develop models for refinery operations that can be used to optimize the various processes and also to predict product yields and properties.

The purpose of this case study is to illustrate how to develop such models. Since the FCC process is the most important and widely used process in a petroleum refinery, the present work is concerned with predicting yields and properties for this process. The consideration of other processes can be undertaken in a similar manner.

2.4.1
Fluid Catalytic Cracking Process

Figure 2.2 shows a simplified process flow chart of the FCC unit. The cracking reactions are carried out in a vertical reactor vessel in which vaporized oil rises and carries along with it, in intimate contact, small fluidized catalyst particles. The reactions are very rapid, and only a few seconds of contact time are necessary for most applications. Simultaneously with the desired reactions, a carbonaceous material of low hydrogen-to-carbon (H/C) ratio, "coke" deposits on the catalyst and renders it inactive for all practical purposes. The spent catalyst and the converted oils are then separated, and the catalyst is passed down-flow to a separate chamber, the regenerator, where the coke is combusted, rejuvenating the catalyst. The regenerated catalyst is then conveyed down-flow to the bottom of the reactor riser, where the cycle begins again.

A number of mechanistic modeling studies to explain the fluid catalytic cracking process and to predict the yields of valuable products of the FCC unit have been performed in the past. Weekman and Nace (1970) presented a reaction network model based on the assumption that the catalytic cracking kinetics are second order with respect to the feed concentration and on a three-lump scheme. The first lump corresponds to the entire charge stock above the gasoline boiling range, the second

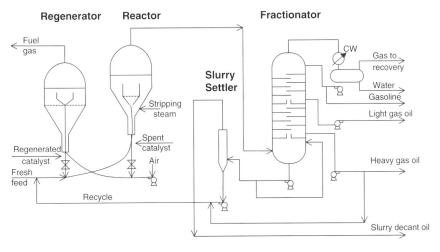

Figure 2.2 Fluid catalytic cracker unit.

lump represents the gasoline range hydrocarbon products, and the third lump corresponds to the coke and C_1–C_4 products. The major advantages of the Weekman and Nace model are its simplicity in describing the cracking reactions and its ability to provide an estimation of gas oil and gasoline yields. The major disadvantage is the lumping of coke with light gases. Despite this shortcoming, the model is currently being used in many FCC simulation studies.

Jacob, Voltz and Weekman (1976) proposed a kinetic-based FCC reaction network consisting of ten kinetic lumps. In this model, the feed is lumped into paraffins, naphthenes, aromatic rings, and aromatic groups in both the heavy and light fractions of the charge stock, see Figure 2.3. The products are divided into two lumps: gasoline, and coke and C_1–C_4 gases. The advantage of this model is that yields of different products can be estimated. The major disadvantage is the complicated

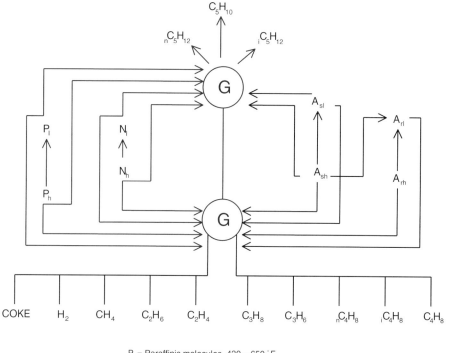

P_l = Paraffinic molecules, 430 – 650 °F
N_l = Naphthenic molecules, 430 – 650 °F
A_{sl} = Carbon atoms among aromatic rings, 430– 650 °F
A_{rl} = Aromatic substituent groups, 430 - 650 °F
P_h = Paraffinic molecules, 650 °F+
N_h = Naphthenic molecules, 650 °F+
A_{sh}= Carbon atoms among aromatic rings, 650 °F+
A_{rh}= Aromatic substituent groups, 650 °F+
G = G lunp (C5 - 430 °F)
C = C lunp (Cl to C4 + Coke)
A_{sl} +P_l + N_l +A_{rt} = LFO (430 – 650°F)
A_{sh}+P_h + N_h+A_{rh}= HFO (650 °F+)

Figure 2.3 Ten-lump FCC reaction network.

mathematics, the necessity of analyzing the feed composition beforehand, and, as in the three-lump model, the lumping of coke and light gases in one component, despite the different role these components play in the behavior of the FCC unit. Arbel *et al.* (1995) presented an extension to the Jacobs *et al.* model that takes into account catalyst characterization based on experimental data.

Models that distinguish between coke and light gases have also appeared in the literature. Lee *et al.* (1989) proposed such a model with a four-lump kinetic network. Their results showed that the calculated yield values for gasoline, coke, and gas oil were consistent with experimental data. Corella and Frances (1991) have also proposed a model that distinguishes between coke and light gases. Their model is based on a five-lump kinetic scheme: feedstock, gas oil, gasoline, light gases, and coke. This scheme suggests that the feedstock cracks to gas oil, which makes it more applicable to hydrocracking units.

From the above discussion, it is clear that the performance of the FCC process depends upon a number of factors such as: feedstock quality, conversion level, and operating conditions. When carrying out performance studies or when comparing competing processes, mechanistic models are usually not the best course of action. Experienced operators have relied in the past on light gas correlations based on industrial yield data (e.g., Gary and Handwerk, 2001; Maples, 1993). These are usually charts that give a general description of light gas yields as a function of feed conversion and feed gravity. In this section we will illustrate the development of linear and non-linear regression models that give the FCC product yields and properties. The feedstock properties are the primary correlation parameters. The models can be based on data collected from the plant itself or from various other sources from the literature on pilot and commercial plants. The main utility of these models is their ease of integration within general mathematical programming models for planning and scheduling of refinery operations. The correlations that we present below have the ability to predict the yield and product properties that a typical FCC unit would achieve in practice. Although the operating conditions have a great effect on the FCC process, these are excluded because they are not really needed to evaluate differences in product yields and properties for different feedstocks.

2.4.2
Development of FCC Process Correlation

In order to develop a general correlation for the FCC process, data was collected from various sources on pilot and commercial plant operations (Maples, 1993; Venuto and Habib, 1979; Vermilion, 1971; Young *et al.*, 1991). In the data, the yield using a zeolite catalyst was given for the following products:

1) Propylene
2) Isobutane
3) Butylene
4) Butane
5) Gasoline

6) Coke
7) Heavy cycle oil
8) Light cycle oil
9) Propane
10) Gas (lighter than C2)
11) Normal butane

The yields for the various products were given as a function of the properties of the feedstocks and the volume percent conversion. The feedstock properties are the feed sulfur content expressed as a weight percent, the feed API and Watson characterization factor, K. The API is related to specific gravity by the following equation:

$$\text{API} = \frac{141.5}{\text{standard specific gravity}} - 131.5$$

where the standard specific gravity (SG) is the ratio of the material density at 60 °F to the water density at the same temperature. The characterization factor K is defined as (Watson and Nelson, 1933):

$$K = \frac{(T_{\text{mabp}})^{\frac{1}{3}}}{SG}$$

where T_{mabp} is the mean average boiling point of the petroleum fraction in degrees Rankin.

The data encompass a wide range of conditions for the FCC process. Table 2.1 gives the range of data collected. Histograms of the yields of the products of the unit and of the carbon Conradson residue (CCR) number are given in Figure 2.4. The histograms are all bell-shaped and illustrate the wide range of data assembled.

Table 2.1 Ranges of data collected.

Variable	Minimum value	Maximum value
Feed K	11.3	12.2
LV%	24.3	93
Feed API	12.9	39.4
Feed S	0.1	6
Propane (C_3) yield	0.6	5.6
n-Butane (nC_4) yield	0.3	3.8
Butane(C_4)	4.6	32.2
Isobutane (iC_4) yield	0.3	14
Propylene ($C_3=$) yield	1.6	12
Butylene ($C_4=$) yield	2	15.2
Light gas yield	0.5	15.4
Gasoline yield	18.2	73
Light cycle oil (LCO) yield	1.1	68
Heavy cycle oil (HCO) yield	1.0	40
Coke yield	1.3	11.7
Carbon Conradson number (CCR)	0	4.2

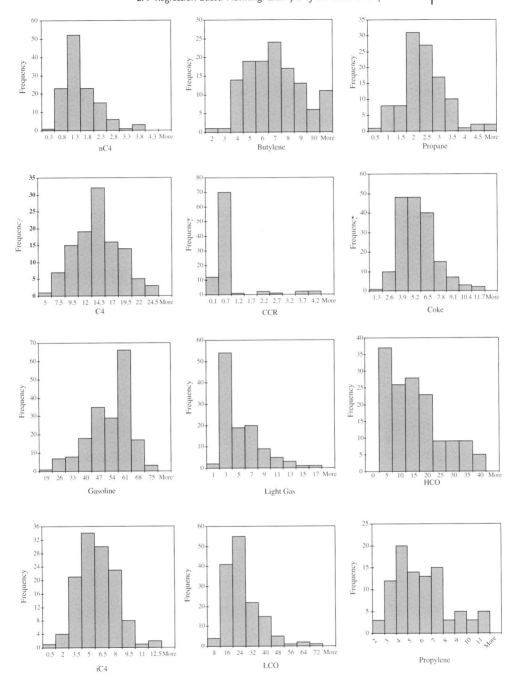

Figure 2.4 Histograms showing the variation of data collected.

2 Petroleum Refinery Planning

To simplify the discussion, let:

X_1 = Feed characterization factor
X_2 = Liquid volume percent conversion
X_3 = Feed API
X_4 = Feed sulfur content

Let also Y_i denote product i yield and Z_{ij} denote property j for product i. The functional relationship for each dependent variable Y_i (or Z_{ij}) can be first assumed to be a linear additive function of the four independent variables, that is,

$$\hat{Y}_i = A_i + B_i X_1 + C_i X_2 + D_i X_3 + E_i X_4 + e_i$$

The error term e_i in the above equation is the deviation of the value of $\hat{Y}_i$ from the true value Y_i. A least square analysis was carried out for each dependent variable $\hat{Y}_i$ with the objective of finding the best linear equation that fits the data with respect to the criteria: minimizing the sum of the error square (i.e., minimize $\sum_{i=1}^{N_d}(Y_i - \hat{Y}_i)^2 = \sum_{i=1}^{N_d} e_i^2$) Where N_d is the number of data points collected for the variable Y_i.

Table 2.2 gives properties of some of the important FCC unit products. All correlations are linear and with a good fit, where the correlation coefficient R is close to 1. In Table 2.3 the yields of the various FCC products are given as linear equations. The utility of these equations is their ease of integration within linear programming (LP) software for a refinery. The correlation coefficient R is also given in Table 2.3. As can be seen, most yield equations have a coefficient that is close to 1. For those equations with a poor linear fit, a nonlinear model would be more appropriate. The procedure to obtain nonlinear functional forms for yield equations with good predictions is a trial and error one. Various forms have been tried that include different nonlinear terms (e.g., cross terms, logs, exponentials, etc.). The obtained correlations are shown in Table 2.4. The predictions of these models are more accurate than those of the linear models. However, their use in refinery planning programs would increase the computation burden. In addition, with a linear program sensitivity and post optimality can be easily carried out; unlike nonlinear programs that require resolving the model.

Improvement in the accuracy of the linear yield equations might be obtained by using more independent variables in the prediction process. For instance, the use of aniline point and volumetric average boiling point (VABP) might lead to better accuracy. Unfortunately, the published data do not include these two variables.

Table 2.2 FCC unit product properties.

Wt.% sulfur in hydrogen sulfide = $-0.0683 + 0.451$ Wt.% in feed	$R = 0.999$
Wt.% sulfur in LCO = $-0.203 + 1.21$ Wt. % in feed	$R = 1$
Wt. % sulfur in HCO = $-0.365 + 2.16$ Wt. % in feed	$R = 1$
Clear Research Octane Number = $87.1 + 0.27 X_2 - 0.27$ (Gaso)	$R = 0.998$
Gasoline API = $64.9 - 0.193 X_2 + 0.0687$ (GASO)	$R = 1$
LCO API = $-3.24 - 0.482 X_2 + 2.15 X_3$	$R = 1$

Table 2.3 Linear regression models for predicting FCC product yields.

(C3=) = $-0.615\ X_1 + 0.208\ X_2 - 0.0465 X_3 + 0.7\ X_4 + 0.07$	$R = 0.923$
(I-C4) = $1.32\ X_1 + 0.155\ X_2 + 0.056\ X_3 - 0.0517\ X_4 - 19.7$	$R = 0.938$
(C4=) = $0.957\ X_1 + 0.166\ X_2 - 0.0916\ X_3 + 1.17\ X_4 - 13.9$	$R = 0.944$
(C4) = $3.43\ X_1 + 0.235\ X_2 + 0.216\ X_3 + 0.443\ X_4 - 47.8$	$R = 0.948$
(GASO) = $-0.66\ X_1 + 0.754\ X_2 - 0.362\ X_3 - 1.333\ X_4 + 19.8$	$R = 0.962$
(COKE) = $1.11\ X_1 + 0.0663\ X_2 - 0.118\ X_3 + 0.185\ X_4 - 9.96$	$R = 0.836$
(HCO) = $-3.13\ X_1 - 0.47\ X_2 + 0.207\ X_3 + 3.75\ X_4 + 71.4$	$R = 0.883$
(LCO) = $1.82\ X_1 - 0.44\ X_2 - 0.018\ X_3 - 2.72\ X_4 + 30$	$R = 0.776$
(GAS) = $0.756\ X_1 - 0.0349\ X_2 + 0.0188\ X_3 - 0.373\ X_4 - 4.31$	$R = 0.738$
(N-C4) = $0.159\ X_1 + 0.0136\ X_2 - 0.00135\ X_3 - 0.0743\ X_4 - 1.57$	$R = 0.785$
(C3) = $-0.74\ X_1 + 0.0175\ X_2 + 0.0598\ X_3 - 0.267\ X_4 + 0.54$	$R = 0.647$

2.4.3
Model Evaluation

The models presented in the previous section are evaluated by comparing their predictions to actual plant data and also to the predictions of correlations proposed by previous researchers. The models give predictions that are close to the actual data. The average relative percent error between the predicted values and the actual values ranges from 0 to 10% for the proposed models. Figure 2.5 represents cross plots of the predicted versus the actual product yields and properties. As can be seen, all data points lie close to the 45° line, indicating a good fit.

The predictions of the proposed correlations are compared to those of Maples (1993), Gary and Handwerk (2001), and Jones (1995). Tables 2.5–2.7 give the percent error for the predictions of the various models. As can be seen, the predictions of the present correlations are always closer to the actual data.

2.4.4
Integration within an LP for a Petroleum Refinery

An LP model for a petroleum refinery would consist of a number of modules for the different units (e.g., FCC, hydrocracker, fluid delayed coker, etc.). Each of these

Table 2.4 Nonlinear regression models for predicting FCC product yields.

LOG(HCO) = $-0.303\ X_1 - 0.0194\ X_2 + 0.0207\ X_3 + 0.0882\ X_4 + 5.38$	$R = 0.952$
(COKE) = $40.9 \text{LOG}\ X_1 + 12.2\ \text{LOG}\ X_2 - 0.0082\ X_3/X_4 - 60.8$	$R = 0.911$
LOG(LCO) = $-0.98\ \text{LOG}\ X_1 - 0.782\ \text{EXP}(0.01 X_2) - 0.00077\ X_3{}^*X_4/X_1 + 3.93$	$R = 0.921$
(GAS) = $1.66\ X_3/X_1 - 2.43\ \text{EXP}(0.01 X_2) - 2.58\ \text{EXP}(0.1 X_4) + 6.54$	$R = 0.951$
LOG(N-C4) = $3.05\ \text{LN}\ X_1 + 0.252\ \text{EXP}\ (0.01 X_2) + 0.184\ \text{EXP}$ $(0.05\ X_3) - 0.891\ \text{EXP}\ (0.095 X_4) - 7.59$	$R = 0.906$
(C3) = $-186\ \text{LOG}\ X_2 - 0.0461\ X_2\ (X_1 + X_3 + X_4)/100 + 0.0747$ $X_2{}^2 - 0.35\ X_2{}^2\ (X_1 + X_3 + X_4)/100 + 5 X_2{}^2 (X_1 + X_3 + X_4)/100)^2)$ $-12((X_1 + X_3 + X_4)/100)^2) + 303$	$R = 0.910$

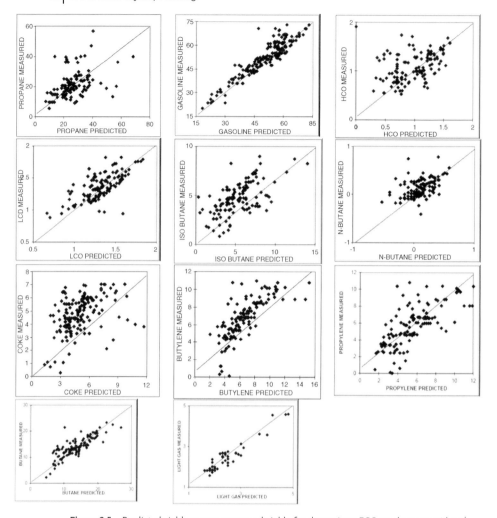

Figure 2.5 Predicted yields versus measured yields for the various FCC products considered.

modules represents the material balances for the unit. The interconnections among units would be appropriately represented by information flows. For instance, to indicate that the product of one unit would be the feed to another unit, the variables are defined in such a way that the information flow is in the same direction as the material flow.

The material balance equations for the FCC unit are easily expressed in terms of the yield equations presented earlier. If F represents the total inlet feed (barrels per day, BPD) to the FCC unit and Y_i is the yield of product i as read from Tables 2.3 and 2.4, then the production of product i can be simply obtained by multiplying the feed to the unit by the yield of product i (i.e., FY_i). Such material balance equations must be written for all units of a refinery in order to prepare a mathematical

Table 2.5 Gasoline comparison yields.

Data point	Feed K	LV%	Feed API	Feed S	Actual gasoline	Gasoline predicted (Gary)	Error%	Gasoline predicted (Jones)	Error%	Gasoline predicted (Maple)	Error%	Gasoline predicted (this work)	Error%
1	12.2	90	27.5	0.5	71.1	66.21	6.878	51.44	27.65	71.81	0.999	68.988	2.97
2	11.6	85	34	0.5	68	70.049	3.013	50.86	25.21	67.575	0.625	63.261	6.969
3	12.2	83.5	27.4	0.6	64.1	62.8559	1.941	50.686	20.93	66.3045	3.439	63.9902	0.171
4	12	80	23.3	0.3	63	57.1008	9.364	50.28	20.19	63.34	0.54	63.3664	0.582
5	12	78	22.5	2.5	59	55.318	6.241	50.048	15.17	61.646	4.485	59.222	0.376
6	11.8	75	24.8	2.2	57.2	56.0598	1.993	49.7	13.11	59.105	3.33	56.6584	0.947
7	11.8	72.3	25.9	0.7	57.8	55.7807	3.494	49.3868	14.56	56.8181	1.699	56.2194	2.735
8	11.5	70	22	0.1	56.3	50.822	9.73	49.12	12.75	54.87	2.54	56.893	1.053
9	12.1	68	19.4	0.3	55.6	47.2824	14.96	48.888	12.07	53.176	4.36	55.6642	0.115
10	11.8	65	24.8	2.2	48.9	51.0498	4.396	48.54	0.736	50.635	3.548	49.1184	0.447
11	11.8	63	31.4	0.5	47.6	56.4894	18.68	48.308	1.487	48.941	2.817	47.4822	0.247
12	11.8	61	23.7	1.6	48.3	47.9722	0.679	48.076	0.464	47.247	2.18	47.2986	2.073
13	11.8	59	21.4	0.5	47.8	44.7254	6.432	47.844	0.092	45.553	4.701	48.0862	0.599
14	11.8	55.7	25.1	0.3	44.8	46.6833	4.204	47.4612	5.94	42.7579	4.558	44.4714	0.733
15	11.8	55	24.8	2.2	41.4	46.0398	11.21	47.38	14.44	42.165	1.848	41.5784	0.431
16	12	45	24.8	2.2	35.9	41.0298	14.29	46.22	28.75	33.695	6.142	33.9064	5.553
17	12.1	42.5	28.8	0.6	31.8	43.6813	37.36	45.93	44.43	31.5775	0.7	32.6354	2.627
18	12	35	39.4	0.3	25.6	50.2694	96.36	45.06	76.02	25.225	1.465	23.6082	7.78
19	12.1	29	31	0.6	22.6	39.065	72.85	44.364	96.3	20.143	10.87	21.66	4.159
20	11.3	24.5	20.3	2.7	18.2	26.3673	44.88	43.842	140.9	16.3315	10.27	19.8754	9.205

34 | 2 Petroleum Refinery Planning

Table 2.6 Propylene comparison yields.

Data point	Feed K	LV%	Feed API	Feed S	Actual propylene	Propylene predicted (Gary)	Error %	Propylene predicted (Jones)	Error %	Propylene predicted (Maple)	Error %	Propylene predicted (this work)	Error %
1	12.1	68	19.4	0.3	5.9	4.0248	31.78	21.152	258.5	6.07	2.8814	6.08	3.06
2	11.5	70	22	0.1	6.8	4.914	27.74	22.18	226.2	6.28	7.6471	6.60	2.88
3	11.5	70	22	0.1	6.7	4.914	26.66	22.18	231	6.28	6.2687	6.60	1.43
4	11.5	70	22	0.1	6.9	4.914	28.78	22.18	221.4	6.28	8.9855	6.60	4.28
5	11.8	60	22.5	2.6	6.1	4.005	34.34	17.04	179.3	5.23	14.262	6.07	0.55
6	11.8	75	24.8	2.2	8.9	6.1676	30.7	24.75	178.1	6.805	23.539	8.80	1.13
7	11.8	75	24.8	2.2	9.1	6.1676	32.22	24.75	172	6.805	25.22	8.80	3.30
8	11.8	65	24.8	2.2	6.8	5.1276	24.59	19.61	188.4	5.755	15.368	6.72	1.18
9	11.8	65	24.8	2.2	6.9	5.1276	25.69	19.61	184.2	5.755	16.594	6.72	2.61
10	11.8	55	24.8	2.2	4.7	4.0876	13.03	14.47	207.9	4.705	0.1064	4.64	1.28
11	11.8	55	24.8	2.2	4.8	4.0876	14.84	14.47	201.5	4.705	1.9792	4.64	3.34
12	11.8	55	24.8	2.2	4.7	4.0876	13.03	14.47	207.9	4.705	0.1064	4.64	1.28
13	11.8	73	24.9	0.6	7	5.965	14.79	23.6192	237.4	6.574	6.0857	7.22	3.11
14	11.8	73	24.9	0.6	7	5.9546	14.93	23.5678	236.7	6.5635	6.2357	7.20	2.81
15	11.8	58	25.1	1.6	4.7	4.4782	4.719	16.012	240.7	5.02	6.8085	4.83	2.76
16	12.2	73	27	0.3	6.9	6.484	6.029	23.465	240.1	6.5425	5.1812	6.60	4.33
17	11.3	85	27.1	0.8	10.2	7.831	23.23	29.9928	194	7.876	22.784	10.14	0.57
18	11.6	85	27.1	0.8	10.4	7.831	24.7	29.9928	188.4	7.876	24.269	9.96	4.26
19	12	49	28.8	1.5	2.5	4.47	78.8	11.1804	347.2	4.033	61.32	2.51	0.38
20	11.8	61	30.2	1.2	5	6.1472	22.94	17.6568	253.1	5.356	7.12	4.98	0.43

Table 2.7 Light gas comparison yields.

Datapoint	Feed K	LV%	Feed API	Feed S	Actual gas	Light gas predicted (Gary)	Error %	Light gas predicted (Jones)	Error %	Light gas predicted (Maple)	Error %	Light gas predicted (this work)	Error %
1	12.2	93	34	0.1	2.5	4.336	73.44	21.265	750.6	2.723	8.92	2.40144	3.9423
2	11.6	85.2	27.1	0.8	2	5.4274	171.4	19.744	887.2	2.4656	23.28	1.9265	3.6751
3	11.9	84.5	27.1	0.8	1.9	5.321	180.1	19.6075	932	2.4425	28.553	1.86847	1.6595
4	12.2	83.5	27.4	0.3	2	5.07	153.5	19.4125	870.6	2.4095	20.475	2.00893	0.4463
5	12	77.6	23.8	0.5	1.8	5.3612	197.8	18.262	914.6	2.2148	23.044	1.84024	2.2354
6	11.8	76.5	28	0.6	2.5	3.808	52.32	18.0475	621.9	2.1785	12.86	2.51739	0.6955
7	11.5	70	22	0.1	2.3	4.8	108.7	16.78	629.6	1.964	14.609	2.2163	3.639
8	12.1	68	27	0.7	2.8	2.846	1.643	16.39	485.4	1.898	32.214	2.68054	4.2665
9	11.8	66.2	22	0.1	2.4	4.2224	75.93	16.039	568.3	1.8386	23.392	2.31803	3.4155
10	11.8	64	27.4	0.3	3	2.106	29.8	15.61	420.3	1.766	41.133	3.12756	4.2518
11	11.8	62.6	22	0.1	2.3	3.6752	59.79	15.337	566.8	1.7198	25.226	2.48461	8.0263
12	11.8	60	22.5	2.6	1.8	3.115	73.06	14.83	723.9	1.634	9.2222	1.93143	7.3014
13	11.8	60	22.5	2.6	2	3.115	55.75	14.83	641.5	1.634	18.3	1.93143	3.4287
14	11.8	60	22.5	2.6	1.8	3.115	73.06	14.83	723.9	1.634	9.2222	1.93143	7.3014
15	12	50.6	22.9	0.6	3.1	1.5542	49.86	12.997	319.3	1.3238	57.297	2.93779	5.2325
16	12	48.4	23.4	2.2	2.9	1.0548	63.63	12.568	333.4	1.2512	56.855	2.61932	9.6786
17	12	36	30.4	0.4	4.7	−3.14	166.8	10.15	116	0.842	82.085	4.57705	2.6159
18	12	32	29.5	0.4	4.8	−3.451	171.9	9.37	95.21	0.71	85.208	4.58912	4.3933
19	12.1	29	28.7	0.4	4.2	−3.643	186.7	8.785	109.2	0.611	85.452	4.54454	8.2034
20	12.1	29	22.5	0.7	3.7	−1.597	143.2	8.785	137.4	0.611	83.486	3.61219	2.3733

programming (planning) model. In addition, other constraints must also be generated, such as capacity constraints, component blending restrictions, quality specifications and demand constraints.

2.5
Artificial-Neural-Network-Based Modeling: Example of Fluid Catalytic Cracker

Refinery operations are complex processes and no thorough model has yet been developed for them. These processes are characterized by large dimensionality and a strong interaction among the process variables. Their modeling usually requires the application of mass and heat transfer, fluid mechanics, thermodynamics, and kinetics. The result is a system of nonlinear, coupled algebraic and/or differential equations. A large number of equations is usually required for their description and many parameters have to be estimated. In addition to the above complications, refinery processes are inherently flexible processes that are able to take a variety of feedstocks and to operate at various conditions to be able to meet seasonal and other changing product demand patterns.

A neural-network-based simulator can overcome the above complications because the network does not rely on exact deterministic models (i.e., based on the physics and chemistry of the system) to describe a process. Rather, artificial neural networks assimilate operating data from an industrial process and learn about the complex relationships existing within the process, even when the input–output information is noisy and imprecise. This ability makes the neural-network concept well suited for modeling complex refinery operations. For a detailed review and introductory material on artificial neural networks, we refer readers to Himmelblau (2008), Kay and Titterington (2000), Baughman and Liu (1995), and Bulsari (1995). We will consider in this section the modeling of the FCC process to illustrate the modeling of refinery operations via artificial neural networks.

2.5.1
Artificial Neural Networks

Artificial neural networks (ANN) are computing tools made up of simple, interconnected processing elements called neurons. The neurons are arranged in layers. The feed-forward network consists of an input layer, one or more hidden layers, and an output layer. ANNs are known to be well suited for assimilating knowledge about complex processes if they are properly subjected to input–output patterns about the process.

In order to develop an ANN model for a refinery unit that is able to predict product yields and properties, one must first decide on the important inputs and outputs of the process. The choice of these inputs and outputs is the most important factor in successfully preparing an ANN model. These inputs and outputs must be chosen by carefully examining plant data. A good expertise in relation to the process is necessary.

Another important factor in developing an ANN model is deciding the architecture of the network. This decision is often made on a trial and error basis. First, a network with one hidden layer only and with a fixed number of neurons is chosen. The predictions of the network are then compared to actual plant data. The number of neurons is increased slowly and the predictions of the network are checked continuously. If the predictions are still poor, then one opts to use one additional hidden layer. For each architecture used, the network is trained to learn the input–output pattern fed to it. Training means the determination of the connection weights among the various neurons. Various algorithms can be used to train a neural network. In this study, the back-propagation algorithm is used. When the neural network is trained with this algorithm, the errors between the estimated outputs from the network and the actual outputs are calculated and propagated backward through the net. These errors are used to update the connection weights among the neurons. This operation is repeated until the network outputs are within a pre-specified tolerance of the actual outputs. More details about the back-propagation algorithm and appropriate references can be found in Elkamel (1998).

The formula used in implementing the back-propagation algorithm is given by:

$$W_t = W_{t-1} + \eta \nabla f(X_t, W_{t-1})(Y_t - f(X_t, W_{t-1})) \qquad t = 1, 2, \ldots \qquad (2.1)$$

The index t in the above equation corresponds to the training instance, X is the vector of input variables, Y is the vector of target variables, η is a learning rate, $f(X^t, W^{t-1})$ is a shorthand notation for network output, and ∇f is the gradient of f with respect to the weights W. Note that the weights in the neural network are adjusted according to the error between the target values and the values predicted by the network. In order to speed up the convergence properties of the back-propagation algorithm a momentum term is used along with a variable learning rate. In addition, the Levenberg–Marquardt algorithm is used to perform the optimization over network weights.

2.5.2
Development of FCC Neural Network Model

In order to develop an ANN model for the FCC process, we use here the same data set as in the previous section (Section 2.4). This data set was divided into two sets, one set for training and one set for testing the neural network. The prepared network model is able to predict the yields of the various FCC products and also the CCR number. During training of the neural network, first, only one hidden layer with five neurons was used. This network did not perform well against a pre-specified tolerance of 10^{-3}. The number of neurons in the hidden layer was therefore increased systematically. It was found that a network of one hidden layer consisting of twenty neurons, as shown in Figure 2.6, performed well for both the training and testing data set. More details about the performance of this network will be given later. The network architecture depicted in Figure 2.6 consists of an input layer, a hidden layer, and an output layer. Each neuron in the input layer corresponds to a particular feed property. The neurons

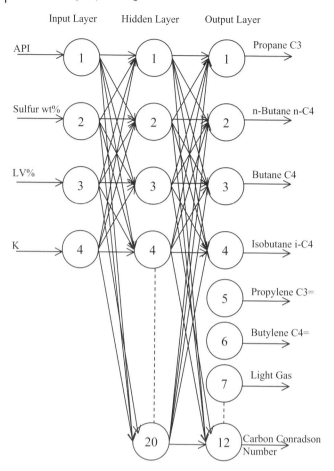

Figure 2.6 Neural network architecture for the FCC model.

in the output layer correspond to the yields of the products of the FCC unit. In addition, there is a neuron corresponding to the CCR number.

The performance of the neural network model is assessed by comparing its predictions with the actual plant measurements. The percent relative errors between the values predicted by the network and the actual values were calculated, as shown in Table 2.8. The average percent error is always less than 4.1% (column 4 of Table 2.8). In the table the maximum and minimum percent errors are also shown (columns 2 and 3, respectively). As can be seen, the predictions of the neural network are always good. The performance of the prepared neural network was also checked against the testing data set that was not used while preparing the network. As can be seen from Table 2.8 (column 7), the average percent error for this set is always less than 6.3% indicating again the good performance of the neural network model for the FCC process. Cross plots for the predicted quantities and the actual quantities for both the

Table 2.8 Evaluation of the FCC ANN in terms of relative error percent.

Network predicted Property	Training data set			Testing data set		
	Maximum error	Minimum error	Average error	Maximum error	Minimum error	Average error
C_3 yield	19.231	0.000	3.020	21.875	0.000	2.208
NC_4 yield	16.674	0.001	2.816	21.447	0.004	6.233
C_4 yield	10.929	0.000	0.630	3.620	0.000	0.041
IC_4 yield	18.745	0.006	3.641	19.856	0.014	6.322
$C_3=$ yield	8.333	0.000	0.321	0.725	0.000	0.053
$C_4=$ yield	14.755	0.000	1.836	11.030	0.000	1.686
Light gas yield	11.907	0.000	1.173	2.799	0.005	0.684
Gasoline yield	9.778	0.002	1.663	4.533	0.011	0.929
LCO yield	13.945	0.001	3.251	12.416	0.053	3.758
HCO yield	19.110	0.000	3.540	15.907	0.011	4.497
Coke yield	27.123	0.000	4.103	18.154	0.013	4.222
CCR	15.811	0.004	1.693	6.585	0.004	2.125

training and testing data sets also confirm the good predictions of the network (i.e., all data lies very close to the 45° line, Figure 2.7).

2.5.3
Comparison with Other Models

The predictions of the neural network model are compared to those of Maples (1993), Gary and Handwerk (2001), Jones (1995), and the regression models of Section 2.4 (see also Al-Enezi, Fawzi and Elkamel (1999)). Gary and Handwerk presented different charts for estimating the yields of the products of the FCC process. The yields are given either in terms of volume percent conversion and °API or volume percent conversion and Watson characterization factor. The Maple charts also give yields of the products of the FCC process in terms of volume percent conversion and °API. The charts provided by Jones are based on the processing of Sassan crude. This is a light 34 °API crude that contains low metals. Tables 2.9 and 2.10 give a comparison of the differentapproaches. In each case, the percent relative error between the predicted value and the actual value is reported. As can be seen, the predictions of the neural network model are consistently closer to the actual values.

To further check the performance of the neural network model, its predictions are compared to those of an existing simulator available at a local refinery. The simulator is based on modeling the FCC unit from first principles. No indications were given, however, on the type of lumping used. The output of the simulator gives the yields of all products of the unit. Typical operating conditions were used in simulating the FCC process. Ten different case studies were considered, see Table 2.11. The results from the FCC simulator were compared to those of the ANN model, see Tables 2.12–2.14.

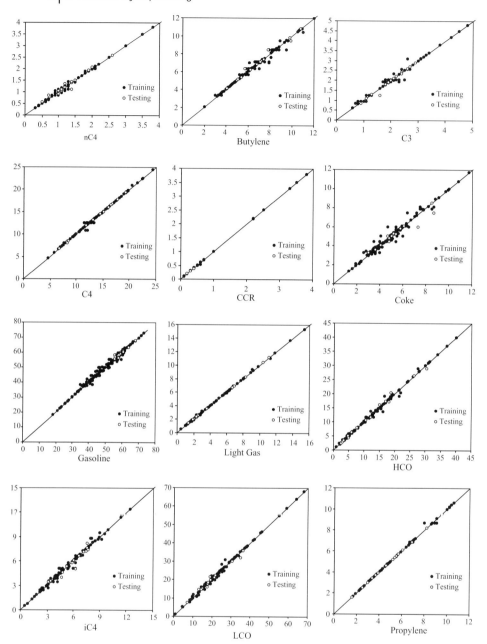

Figure 2.7 Predicted product yield and CCR number versus measured values for the FCC process.

As can be seen from the tables, the ANN model consistently gives better predictions. The main reason is that the simulator required a lot of input information which had to be estimated while the neural network model required only four feed properties.

2.5 Artificial-Neural-Network-Based Modeling: Example of Fluid Catalytic Cracker

Table 2.9 Gasoline comparison yields.

Data point	Feed K	LV%	Feed API	Feed S	Actual gasoline	Gasoline predicted (Gary)	Error %	Gasoline predicted (Jones)	Error %	Gasoline predicted (Maple)	Error %	Gasoline predicted (correlation)	Error %	Gasoline predicted (NN)	% Error
1	12.2	90	27.5	0.5	71.1	66.21	6.88	51.44	27.65	71.81	0.99	68.99	2.97	57.93	18.52
2	11.6	85	34	0.5	68	70.05	3.01	50.86	25.21	67.58	0.63	63.26	6.97	67.99	0.004
3	12.2	83.5	27.4	0.6	64.1	62.86	1.94	50.69	20.93	66.30	3.44	63.99	0.17	64.09	0.0028
4	12	80	23.3	0.3	63	57.10	9.36	50.28	20.19	63.34	0.54	63.37	0.58	62.99	0.004
5	12	78	22.5	2.5	59	55.32	6.24	50.05	15.17	61.65	4.48	59.22	0.38	59.00	0.001
6	11.8	75	24.8	2.2	57.2	56.06	1.99	49.70	13.11	59.12	3.33	56.66	0.95	55.23	3.44
7	11.8	72.3	25.9	0.7	57.8	55.79	3.49	49.39	14.56	56.82	1.70	56.22	2.74	57.79	0.02
8	11.5	70	22	0.1	56.3	50.82	9.73	49.12	12.75	54.87	2.54	56.89	1.05	58.01	3.04
9	12.1	68	19.4	0.3	55.6	47.28	14.96	48.89	12.07	53.18	4.36	55.66	0.12	55.59	0.0008
10	11.8	65	24.8	2.2	48.9	51.05	4.40	48.54	0.74	50.64	3.55	49.12	0.45	49.71	1.66
11	11.8	63	31.4	0.5	47.6	56.49	18.68	48.31	1.49	48.64	2.82	47.48	0.25	47.57	0.07
12	11.8	61	23.7	1.6	48.3	47.97	0.68	48.08	0.46	47.25	2.18	47.30	2.07	48.29	0.0017
13	11.8	59	21.4	0.5	47.8	44.73	6.43	47.84	0.09	45.55	4.70	48.09	0.60	47.79	0.01
14	11.8	55.7	25.1	0.3	44.8	46.68	4.20	47.46	5.94	42.76	4.56	44.47	0.73	44.81	0.02
15	11.8	55	24.8	2.2	41.4	46.04	11.21	47.38	14.44	42.17	1.85	41.58	0.43	44.36	7.15
16	12	45	24.8	2.2	35.9	41.03	14.29	46.22	28.75	33.69	6.14	33.91	5.55	37.52	4.52
17	12.1	42.5	28.8	0.6	31.8	43.68	37.36	45.93	44.43	31.58	0.70	32.64	2.63	31.80	0.004
18	12	35	39.4	0.3	25.6	50.27	96.36	45.06	76.02	25.23	1.46	23.61	7.78	25.60	0.0007
19	12.1	29	31	0.6	22.6	39.06	72.85	44.36	96.3	20.14	10.87	21.66	4.16	22.60	0.0006
20	11.3	24.5	20.3	2.7	18.2	26.3673	44.88	43.842	140.9	16.3315	10.27	19.8754	9.205	18.198	0.0109

Table 2.10 Propylene comparison yield.

Data point	Feed K	LV%	Feed API	Feed S	Actual propylere	Propylene predicted (Gary)	Error %	Propylene predicted (Jones)	Error %	Propylene predicted (Maple)	Error %	Propylene predicted (correlation)	Error %	Propylene predicted (NN)	Error %
1	12.1	68	19.4	0.3	5.9	4.0248	31.78	21.152	258.5	6.07	2.8814	6.08	3.06	5.8999	0.001
2	11.5	70	22	0.1	6.9	4.914	27.74	22.18	226.2	6.28	7.6471	6.60	2.88	6.85	0.7252
3	11.5	70	22	0.1	6.7	4.914	26.66	22.18	231	6.28	6.2687	6.60	4.28	6.85	2.2382
4	11.8	60	22.5	2.6	6.1	4.005	34.34	17.04	179.3	5.23	14.262	6.07	0.55	6.10004	0.0006
5	11.8	75	24.8	2.2	8.9	6.1676	30.7	24.75	178.1	6.805	23.539	8.80	1.13	8.6667	2.6219
6	11.8	75	24.8	2.2	9.1	6.1676	32.22	24.75	172	6.805	25.22	8.80	3.30	8.6667	4.7621
7	11.8	65	24.8	2.2	6.8	5.1276	24.59	19.61	188.4	5.755	15.368	6.72	1.18	6.85	0.7347
8	11.8	65	24.8	2.2	6.9	5.1276	25.69	19.61	184.2	5.755	16.594	6.72	2.61	6.85	0.7241
9	11.8	55	24.8	2.2	4.7	4.0876	13.03	14.47	207.9	4.705	0.1064	4.64	1.28	4.7499	1.0615
10	11.8	73	24.9	0.6	7	5.965	14.79	23.6192	237.4	6.574	6.0857	7.22	2.81	6.9834	0.2369
11	11.8	58	25.1	1.6	4.7	4.4782	4.719	16.012	240.7	5.02	6.8085	4.83	2.76	7.0167	0.2384
12	12.2	73	27	0.3	6.9	6.484	6.029	23.465	240.1	6.5425	5.1812	6.60	4.33	4.7003	0.0063
13	11.3	85	27.1	0.8	10.2	7.831	23.23	29.9928	194	7.876	22.784	10.14	0.57	6.9	0
14	11.6	85	27.1	0.8	10.4	7.831	24.7	29.9928	188.4	7.876	24.269	9.96	4.26	10.1999	0.0006
15	12	49	28.8	1.5	2.5	4.47	78.8	11.1804	347.2	4.033	61.32	2.51	0.38	10.3999	0.0006
16	11.8	61	30.2	1.2	5	6.1472	22.94	17.6568	253.1	5.356	7.12	4.98	0.43	2.4999	0.0003

Table 2.11 Simulation case studies.

Ser. no.	FDK	CONV	FD API	FD S
1	11.5	70	22	0.1
2	11.8	65.2	29.6	0.4
3	11.8	62.6	22	0.1
4	11.8	61.6	22	0.1
5	11.8	59.1	30.7	0.8
6	11.8	56	29.6	0.4
7	11.8	55.9	28.5	0.4
8	11.8	55.5	29.6	0.4
9	11.8	54.5	25.1	0.3
10	12	45.2	25.1	0.4

The artificial neural network model of the FCC gives good predictions of the product yields of the process. The feed properties, feed °API, Watson characterization factor K, feed sulfur content, and liquid volume percent conversion LV%, are the only required inputs. This network can contribute significantly to enhancing the performance of the unit and can be used for selection of feedstocks. The demand for products of the FCC unit is seasonal and fluctuates very often. The network model can be used to properly plan the operation of a refinery according to fluctuating demand. In addition, the prepared model can be optimized to determine the blend of feedstocks appropriate for yielding a pre-specified product distribution. This can be done by invoking the inverse property of artificial neural networks. Once the network has been trained properly and the connecting weights among neurons are obtained, a nonlinear optimization can be carried out to obtain feedstock properties that lead to the desired blend of production. The neural network model can also be used to train

Table 2.12 Comparison of ANN models and simulation results for coke and gasoline yields.

Ser. no.	Coke measured	Coke predicted (simulator)	Coke predicted (ANN)	Gaso. measured	Gaso. predicted (simulator)	Gaso predicted (ANN)
1	6	4.6287	5.02	56.3	66.5211	58.01
2	5.1	4.3881	5.11	46.6	58.2994	46.59
3	4.1	4.6288	4.09	53	66.5221	53.49
4	4.1	4.6261	4.06	51.1	66.5201	51.08
5	3.9	4.4525	3.84	41.9	58.0339	41.21
6	3.9	4.3897	3.7	39.1	58.2996	39.23
7	3.9	4.3689	3.89	40.8	58.2295	40.81
8	3.9	4.3888	3.91	40.5	58.2968	40.21
9	2.9	4.4697	3.02	43.5	61.7550	43.49
10	3.6	4.4767	3.6	33.8	61.6403	33.80

2 Petroleum Refinery Planning

Table 2.13 Comparison of ANN models and simulation results for coke and gasoline.

Ser. no.	C_3 measured	C_3 predicted (simulator)	C_3 predicted (ANN)	IC_4 measured	IC_4 predicted (simulator)	IC_4 predicted (ANN)
1	2.5	3.6629	2.4	6.3	6.9124	5.1
2	2.9	3.9543	2.9	8.8	7.6583	8.68
3	1.7	3.6633	1.69	4.0	6.9129	4.13
4	1.8	3.6620	1.81	4.6	6.9106	4.48
5	2.1	4.2086	2.10	9.5	7.9318	9.34
6	2.1	3.9550	2.09	7.1	7.6594	6.67
7	2.0	3.6664	2.00	8.0	7.3335	8.80
8	1.9	3.9530	1.92	7.5	7.6575	7.17
9	2.8	3.5632	2.80	6.1	7.0456	6.05
10	2.3	3.5529	2.30	4.2	7.0357	4.18

new operators of the FCC unit and illustrate to them the effect of changing feedstocks on the outcome of the unit.

The major advantage of the prepared neural network FCC model is that it does not require a lot of input information. In addition, the model can be updated whenever new input–output information for the FCC unit is made available. This can be done by retraining the neural network starting from the old connection weights as an initial guess for the optimization process and by including the new set of data within the overall set used to train the network.

2.6 Yield Based Planning: Example of a Single Refinery

In this section, we use a simplified refinery model in order to demonstrate the concept of using mathematical programming for the planning of process-yield-based

Table 2.14 Comparison of ANN models and simulation results for n-butane and butylene yields.

Ser. no.	NC_4 measured	NC_4 predicted (simulator)	NC_4 predicted (ANN)	$C_4=$ measured	$C_4=$ predicted (simulator)	$C_4=$ predicted (ANN)
1	1.4	1.9785	1.10	5.7	3.0975	6.33
2	2.6	2.1920	2.60	5.7	3.4311	5.70
3	1.0	1.9786	0.92	5.5	3.0977	5.50
4	1.1	1.9779	1.16	5.9	3.0967	5.90
5	2.2	2.2706	2.20	5.0	3.5538	5.00
6	1.9	2.1924	1.94	5.4	3.4316	5.39
7	2.2	2.0988	2.20	4.2	3.2853	4.20
8	2.1	2.1918	2.00	4.7	3.4307	4.74
9	1.5	2.0163	1.50	3.8	3.1566	3.80
10	1.1	2.0134	1.10	4.2	3.1521	4.20

2.6 Yield Based Planning: Example of a Single Refinery

models. We use the refinery production planning example proposed by Allen (1971). This example illustrates the planning of petroleum refinery operations as an ordinary single objective linear programming (LP) problem of total daily profit maximization.

Figure 2.8 is a block diagram representation of a refinery that is essentially made up of a primary crude distillation unit (CDU) and a middle distillates cracker, widely known as a catalytic cracker in modern settings. The refinery processes crude oil to produce gasoline, naphtha, jet fuel, heating oil, and fuel oil. The primary unit splits the crude into naphtha (13 wt% yield), jet fuel (15 wt%), gas oil (22 wt%), cracker feed (20 wt%), and residue (30 wt%). Gasoline is blended from naphtha and cracked blend stock in equal proportions. Naphtha and jet fuel products are straight run. Heating oil is a blend of 75% gas oil and 25% cracked oil. Fuel oil can be blended from primary residue, cracked feed, gas oil, and cracked oil, in any proportions. The process yields of the cracker unit are flared gas (5 wt%), gasoline blend stock (40 wt%) and cracked oil (55 wt%). All the variables are in the same units of ton/day, symbolically denoted as t/day, and are first assigned to process streams flowrate, see Figure 2.8. Since, in linear programming, decision variables cannot feasibly be negative, assigning a variable to a stream also defines its direction of flow and prevents the possibility of reverse flow.

The minimum number needed to define a system fully should be identified. In this example, it is three since, for example, fixing x_1 determines x_7, x_4, x_8, x_9, and x_{10}; fixing x_2 then determines x_{11}, x_{16}, x_3, x_{14}, x_{17}, x_{20}, and x_{15}; and finally fixing x_5 then determines x_{12}, x_{18}, x_{13}, x_{19}, and x_6. An LP model could be formulated using only these three structural variables or any other suitable three variables. In this case, the solution would only give values of the three variables and the remainder (if needed) would have to be calculated from them afterwards. It is usually more convenient to include additional variables in LP programs.

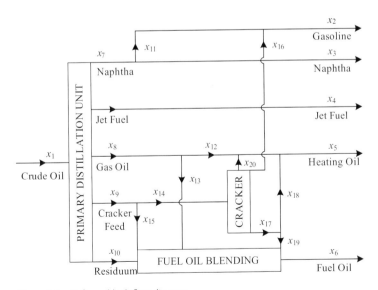

Figure 2.8 Refinery block flow diagram.

2.6.1
Model Formulation

In this section we construct the mathematical formulation that describes the above refinery.

2.6.1.1 Limitations on Plant Capacity

In this example, the feed rates of crude oil to the primary unit and cracker, averaged over a period of time, can be anything from zero to the maximum plant capacity. The constraints are:

$$\text{primary distillation unit:} \quad x_1 \leq 15\,000 \tag{2.2}$$

$$\text{cracker:} \quad x_{14} \leq 2500 \tag{2.3}$$

2.6.1.2 Material Balances

Material balance constraints are in the form of equalities. There are three types of such constraints: fixed plant yield, fixed blends or splits, and unrestricted balances. Except in some special situations, such as planned shutdown of the plant or storage movements, the right hand-side of the balance constraints is always zero. For the purpose of consistency, flow into the plant or stream junction has negative coefficients and flows out have positive coefficients. The constraints are as follows:

Fixed Plant Yield For the primary distillation unit:

$$-0.13x_1 + x_7 = 0 \tag{2.4}$$

$$-0.15x_1 + x_4 = 0 \tag{2.5}$$

$$-0.22x_1 + x_8 = 0 \tag{2.6}$$

$$-0.20x_1 + x_9 = 0 \tag{2.7}$$

$$-0.30x_1 + x_{10} = 0 \tag{2.8}$$

For the cracker:

$$-0.05x_{14} + x_{20} = 0 \tag{2.9}$$

$$-0.40x_{14} + x_{16} = 0 \tag{2.10}$$

$$-0.55x_{14} + x_{17} = 0 \tag{2.11}$$

Fixed Blends For gasoline blending:

$$0.5x_2 - x_{11} = 0 \tag{2.12}$$

$$0.5x_2 - x_{16} = 0 \tag{2.13}$$

For heating oil blending:

$$0.75x_5 - x_{12} = 0 \tag{2.14}$$

$$0.25x_5 - x_{18} = 0 \tag{2.15}$$

Unrestricted Balances

Naphtha : $\quad -x_7 + x_3 + x_{11} = 0 \tag{2.16}$

Gas oil : $\quad -x_8 + x_{12} + x_{13} = 0 \tag{2.17}$

Cracker feed : $\quad -x_9 + x_{14} + x_{15} = 0 \tag{2.18}$

Cracked oil : $\quad -x_{17} + x_{18} + x_{19} = 0 \tag{2.19}$

Fuel oil : $\quad -x_{10} - x_{13} - x_{15} - x_{19} + x_6 = 0 \tag{2.20}$

2.6.1.3 Raw Material Limitation and Market Requirement

The constraints considered so far are concerned with the physical nature of the plant. Constraints are also needed to define the availability of raw materials and product requirements over a time period. For this example, there are no restrictions on crude oil availability or the minimum production required. The maximum production requirement constraints, in t/day, are as follows:

Gasoline : $\quad x_2 \leq 2700 \tag{2.21}$

Naphtha : $\quad x_3 \leq 1100 \tag{2.22}$

Jet fuel : $\quad x_4 \leq 2300 \tag{2.23}$

Heating oil : $\quad x_5 \leq 1700 \tag{2.24}$

Fuel oil : $\quad x_6 \leq 9500 \tag{2.25}$

Since linear programming variables cannot feasibly be negative, an additional constraint to be specified is:

$$x_1, x_2, \ldots, x_{20} \geq 0 \text{ or } x_i \geq 0, \ i = 1, 2, \ldots 20 \tag{2.26}$$

2.6.1.4 Objective Function

The objective function can assume different representation with regards to the system under study. A commonly used objective of an industrial process is to maximize profit or to minimize the overall costs. The former is adopted in this work. In this model, the whole refinery is considered to be one process, where the process uses a given petroleum crude to produce various products in order to achieve specific economic objectives. Thus, the objective of the optimization at hand is to achieve maximum profitability given the type of crude oil and the refinery facilities. No major hardware change in the current facilities is considered in this problem. The

2 Petroleum Refinery Planning

cost of acquiring the raw material crude oil and transforming it to finished products is subtracted from the gross revenues accruing from the sale of finished products. The sign convention denotes costs as negative and realization from sales as positive. Each element consists of the product of coefficient of unit cost or unit sales price ($/ton) and a production flowrate variable (t/day). The objective function is as follows:

$$\underset{x_i}{\text{maximize}}\ z = -8.0x_1 + 18.5x_2 + 8.0x_3 + 12.5x_4 + 14.5x_5 + 6.0x_6 - 1.5x_{14}$$

(2.27)

2.6.2
Model Solution

The deterministic LP model was set up on GAMS and solved using CPLEX. Table 2.15 illustrates the computational results for the refinery Model. The planning model suggested producing 2000 t/day of gasoline, 625 t/day of naphtha, 1875 t/day of jet

Table 2.15 Computational results for deterministic model from GAMS/CPLEX.

Decision variable	Value (t/day)			Dual price/ marginal ($/ton)	Slack or surplus variable	Value (t/day)	Dual price/ marginal ($/ton)
	Lower limit	Level	Upper limit				
x_1	0	12 500	$+\infty$	0	s_1	2 500	0
x_2	0	2 000	$+\infty$	0	s_2	0	3.575
x_3	0	625	$+\infty$	0	s_3	700	0
x_4	0	1 875	$+\infty$	0	s_4	475	0
x_5	0	1 700	$+\infty$	0	s_5	425	0
x_6	0	6 175	$+\infty$	0	s_6	0	8.5
x_7	0	1 625	$+\infty$	0	s_7	3 325	0
x_8	0	2 750	$+\infty$	0	a_8	0	8.0
x_9	0	2 500	$+\infty$	0	a_9	0	12.5
x_{10}	0	3 750	$+\infty$	0	a_{10}	0	6.0
x_{11}	0	1 000	$+\infty$	0	a_{11}	0	9.825
x_{12}	0	1 275	$+\infty$	0	a_{12}	0	6.0
x_{13}	0	1 475	$+\infty$	0	a_{13}	0	0
x_{14}	0	2 500	$+\infty$	0	a_{14}	0	29.0
x_{15}	0	0	$+\infty$	-3.825	a_{15}	0	6.0
x_{16}	0	1 000	$+\infty$	0	a_{16}	0	8.0
x_{17}	0	1 375	$+\infty$	0	a_{17}	0	29.0
x_{18}	0	425	$+\infty$	0	a_{18}	0	6.0
x_{19}	0	950	$+\infty$	0	a_{19}	0	6.0
x_{20}	0	125	$+\infty$	0	a_{20}	0	8.0
					a_{21}	0	6.0
					a_{22}	0	9.825
					a_{23}	0	6.0
					a_{24}	0	6.0
Total Profit ($/day)		23 387.50					

fuel, 1700 t/day of heating oil and 6175 t/day of fuel oil. The total profit associated with this plan was 23 387.50 US$/day.

2.6.3
Sensitivity Analysis

In this section, we will illustrate the concept of sensitivity analysis on both objective function coefficients and right-hand side parameters. In general, sensitivity analysis for the objective function coefficients determines the lower and upper limits of each coefficient that maintains the original optimal solution. Table 2.16 displays the results of these limits in which the original solution or "basis" remains optimal. The allowable increase column indicates the amount by which an objective function coefficient can be increased with the current basis remaining optimal. Conversely, the allowable decrease column is the amount by which the objective coefficient can be decreased with the current basis remaining optimal. We observe that nine out of the 20 decision variables have positive infinity as an upper limit for their associated coefficients. This is reasonable since all decision variables are positive and since the objective is a profit maximization. For example, decision variable x_2 denoting the production flowrate of gasoline has an objective function coefficient value of $18.50/

Table 2.16 Sensitivity analysis for the objective function coefficients.

Decision variable	Objective function coefficient ranges ($/ton)				
	Allowable decrease	Lower Limit	Current value	Upper limit	Allowable increase
x_1	0.715 000	−8.715 000	−8.000 000	−7.235 000	0.765 000
x_2	4.468 750	14.031 25	18.500 000	+∞	+∞
x_3	14.300 000	−6.300 000	8.000 000	13.884 615	5.884 615
x_4	4.766 667	7.733 333	12.500 000	17.600 000	5.100 000
x_5	8.500 000	6.000 000	14.500 000	+∞	+∞
x_6	1.134 921	4.865 079	6.000 000	7.062 500	1.062 500
x_7	5.500 000	−5.500 000	0.000 000	5.884 615	5.884 615
x_8	3.250 000	−3.250 000	0.000 000	3.477 273	3.477 273
x_9	3.575 000	−3.575 000	0.000 000	3.825 000	3.825 000
x_{10}	2.383 333	−2.383 333	0.000 000	2.550 000	2.550 000
x_{11}	8.937 500	−8.937 500	0.000 000	+∞	+∞
x_{12}	11.333 333	11.333 333	0.000 000	+∞	+∞
x_{13}	3.250 000	3.250 000	0.000 000	3.477 273	3.477 273
x_{14}	3.575 000	−5.075 000	−1.500 000	+∞	∞
x_{15}	+∞	−∞	0.000 000	3.825 000	3.825 000
x_{16}	8.937 500	−8.937 500	0.000 000	+∞	+∞
x_{17}	6.500 000	−6.500 000	0.000 000	+∞	+∞
x_{18}	34.000 000	−34.000 000	0.000 000	+∞	+∞
x_{19}	6.500 000	−6.500 000	0.000 000	34.000 000	34.000 000
x_{20}	71.500 000	−71.500 000	0.000 000	+∞	+∞

ton. The price of gasoline can be as low as $4.47/ton without changing the solution basis. The knowledge of such limits extends the refinery flexibility to negotiate prices so long as it is within the bounds of each coefficient, as determined in Table 2.16, especially in the volatile market of spot trading of crude oil and the commodities.

The sensitivity analysis for the right-hand side of constraints was studied, as shown in Table 2.17. The upper limits for some of the constraints are positive infinity while the lower limits vary. Lower limits for most of the constraints assumed a negative value. Constraints whose upper limits are allowed to go to positive infinity imply that they are not critical to the production process. As an illustration, the demand for heating oil can be decreased without inflicting a change in the current optimal basis. The other constraints can be analyzed in a similar manner.

The values of the slack or surplus variables and the dual prices in Table 2.15 provide the most economical average operating plan for a 30-day period. For instance, it indicates that the primary distillation unit is not at full capacity as the solution

Table 2.17 Sensitivity analysis for right-hand side of constraints of deterministic model.

Constraints	Right-hand side of constraints ranges (t/day)				
	Allowable decrease	Lower limit	Current value	Upper limit	Allowable Increase
(2.1)	2 500.000 000	12 500.000 000	15 000.000 000	$+\infty$	$+\infty$
(2.2)	1 340.909 058	1 159.090 94	2 500.000 000	3 000.000 000	500.000 000
(2.3)	625.000 000	−625.000 000	0.000 000	475.000 000	475.000 000
(2.4)	1 875.000 000	−1 875.000 000	0.000 000	425.000 000	425.000 000
(2.5)	1 475.000 000	−1 475.000 000	0.000 000	3 325.000 000	3 325.000 000
(2.6)	500.000 000	−500.000 000	0.000 000	961.538 510	961.538 513
(2.7)	3 750.000 000	−3 750.000 000	0.000 000	3 325.000 000	3 325.000 000
(2.8)	125.000 000	−125.000 000	0.000 000	$+\infty$	$+\infty$
(2.9)	475.000 000	−475.000 000	0.000 000	350.000 000	350.000 000
(2.10)	950.000 000	−950.000 000	0.000 000	3 325.000 000	3 325.000 000
(2.11)	625.000 000	−625.000 000	0.000 000	475.000 000	475.000 000
(2.12)	475.000 000	−475.000 000	0.000 000	350.000 000	350.000 000
(2.13)	1 475.000 000	−1 475.000 000	0.000 000	1 275.000 000	1 275.000 000
(2.14)	950.000 000	−950.000 000	0.000 000	425.000 000	425.000 000
(2.15)	625.000 000	−625.000 000	0.000 000	475.000 000	475.000 000
(2.16)	1 475.000 000	−1 475.000 000	0.000 000	3 325.000 000	3 325.000 000
(2.17)	500.000 000	−500.000 000	0.000 000	961.538 510	961.538 513
(2.18)	950.000 000	−950.000 000	0.000 000	3 325.000 000	3 325.000 000
(2.19)	6 175.000 000	−6 175.000 000	0.000 000	3 325.000 000	3 325.000 000
(2.20)	700.000 000	2 000.000 000	2 700.000 000	$+\infty$	$+\infty$
(2.21)	475.000 000	625.000 000	1 100.000 000	$+\infty$	$+\infty$
(2.22)	425.000 000	1 875.000 000	2 300.000 000	$+\infty$	$+\infty$
(2.23)	1 700.000 000	0.000 000	1 700.000 000	3666.6666	1 966.666 626
(2.24)	3 325.000 000	6 175.000 000	9 500.000 000	$+\infty$	$+\infty$

generates a production mass of 12 500 t/day against its maximum production capacity of 17 300 t/day (as given by the summation of the right-hand-side values of constraints (2.21)–(2.25). Another analytical observation reveals that the maximum production requirement is only met for heating oil.

By definition, the dual price or "shadow price" of a constraint of a linear programming model is the amount by which the optimal value of the objective function is improved (increased in a maximization problem and decreased in a minimization problem) if the right-hand-side of a constraint is increased by one unit, with the current basis remaining optimal. A positive dual price means that increasing the right-hand side in question will improve the objective function value. A negative dual price means that increasing the right-hand side will have a reverse effect. Thus, the dual price of a slack variable corresponds to the effect of a marginal change in the right-hand-side of the appropriate constraint (Winston and Venkataramanan, 2003).

The dual prices of slacks on mass balance and product requirement rows can be interpreted more specifically. Consider a mass balance constraint:

$$-x_1 - x_2 + x_3 = 0 \tag{2.28}$$

where x_3 is the product stream. Introducing the artificial slack variable a_n and then rearranging, we obtain:

$$\begin{aligned} -x_1 - x_2 + x_3 + a_n &= 0 \\ x_1 + x_2 &= x_3 + a_n \end{aligned} \tag{2.29}$$

The product stream is increased by a_n and the feed streams x_1 and x_2 must increase correspondingly. The dual price of a_n indicates the effect of making marginally more products without taking into account its realization (which is on x_3), that is, it indicates the cost added by producing one extra item of the product, or, in other words, the marginal cost of making the product. In addition to that, consider the product requirement constraint:

$$\begin{aligned} x_3 &\leq 1100 \\ \Rightarrow x_3 + s_m &= 1100 \\ \Rightarrow x_3 &= 1100 - s_m \end{aligned} \tag{2.30}$$

The dual price of the slack variable s_m on this constraint indicates the effect of selling this product at the margin, that is, it indicates the marginal profit on the product. If the constraint is slack, so that the slack variable is positive (basic), the profit at the margin must obviously be zero and this is in line with the zero dual price of all basic variables. Since cost + profit = realization for a product, the sum of the dual prices on its balance and requirement constraints equals its coefficient in the original objective function.

In this problem, there are two balance constraints on heating oil, as given by Equations 2.14 and 2.15, the dual prices of which are both $6.00/ton. This is the marginal cost of diverting gas oil and cracked oil from fuel oil to heating oil. The dual price for the constraint on heating oil production as given by inequality (2.24) is $8.50/ton and this is the marginal profit on heating oil, in line with the realization of $14.50/ton in the objective function, as given by the coefficient of x_5.

2.7
General Remarks

The planning and utilization of production capacity is one of the most important responsibilities for managers in the manufacturing industry in general and petroleum refineries in particular. Planning of petroleum refineries typically encompasses different areas, including crude oil management, process unit optimization, and product blending. Crude management entails crude segregation and crude unit operation. Process unit optimization deals with downstream (of the crude distillation unit (CDU)) process unit operations that handle crude unit intermediates. Product blending handles the development of a product shipment schedule and an optimum blend recipe based on information from process unit optimization and current operating data. A major problem in refinery planning is prevalent even at the very foundation: optimization of the CDU and its associated product yields. In addition to uncertainty surrounding the future price of crude oils, the actual composition of crude oils (or crudes, for short) is often only an educated guess. Crudes vary from shipment to shipment because of the mixture of sources actually shipped. It is expected that the quality of crudes does not change significantly over a short period of time, although this assumption could also render a plan to be inaccurate or, worse, infeasible. If the actual crude composition does not closely agree with that modeled, then an error is committed that often propagates through the rest of a refinery planning model.

An equally common source of error in optimizing the submodel for the CDU is the assumption that the fractions from the distillation curve for the crude unit, simply referred to as the swing cuts of distillates, are produced as modeled. Frequently, in practice, models are not even adjusted to show cut overlaps, just because of wishing to take the easy way out in developing crude cut yields and distillates. One of the typical crude cutting procedures assigns distillation temperatures directly from the true boiling point crude analysis, in which no adjustment is made for the actual refinery degree of fractionation. This is a particularly bad procedure for certain types of gasoline that have tight 90% point limits. The fractionation efficiency of gasoline and distillate components from all processes would have a significant effect in controlling aromatics and other types of hydrocarbons. Therefore, planners and decision makers ought to be more diligent by constantly reviewing the supposed optimized plans and comparing them to actual situations in an effort to improve the prediction accuracy of their models.

The third component of the refinery planning involves product blending, where this is usually handled by preparing both a short-range and a long-range plan, using the same model for the blending process. The long-range plan, typically covering 30 days, provides aggregate pools of products for a production schedule. The short-range plan, typically spanning seven days, fixes the blend schedule and creates recipes for the blender. Desired output from the long-range model includes (i) detailed product blend schedule; (ii) optimal blend recipes; (iii) predicted properties of blend recipes; (iv) product and component inventories; (v) component qualities, rundown rates, and costs; (vi) product prices; and (vii) equipment limits. For the short-range model, the desired outputs are: (i) a detailed product blend schedule; (ii) optimal blend recipes; (iii) predicted properties of blend recipes; and (iv) product and component inventories as a function of time (Fisher and Zellhart, 1995).

References

Al-Enezi, G., Fawzi, N., and Elkamel, A. (1999) Development of regression models to control product yields and properties of the fluid catalytic cracking process. *Petroleum Science & Technology*, **17**, 535.

Allen, D.H. (1971) Linear programming models for plant operations planning. *British Chemical Engineering*, **16**, 685.

Anabtawi, J.A., Ali, S.A., and Ali, M.A. (1996) Impact of gasoline and diesel specifications on the refining industry. *Energy Sources*, **18**, 200–214.

Arbel, A., Huang, Z., Rinard, I.H., Shinnar, R., and Sapre, A.V. (1995) Dynamics and control of fluid catalytic crackers: 1. Modelling of the current generation of FCC's. *Industrial Engineering & Chemistry Research*, **34**, 1228.

Baughman, D.R. and Liu, Y.A. (1995) *Neural Networks in Bioprocessing and Chemical Engineering*, Academic Press, San Diego, CA.

Blazek, U. (1971) *Oil & Gas Journal*, **8**, 66.

Bodington, C.E. (1995) Planning, Scheduling, and control Integration in the Process Industries, McGraw Hill, New York.

Bulsari, A.B. (1995) *Neural Networks for Chemical Engineers*, Elsevier, Amsterdam.

Corella, J. and Frances, E. (1991) Modeling some revampings of the riser reactors of the FCCUs. AIChE Annual conference, Los Angeles, USA.

Elkamel, A. (1998) An artificial neural network for predicating and optimizing immiscible flood performance in heterogeneous reservoirs. *Computers & Chemical Engineering*, **22**, 1699.

Elkamel, A., Al-Ajmi, A., and Fahim, M. (1999) Modeling the hydrocracking process using artificial neural networks. *Petroleum Science & Technology*, **17**, 931.

Fisher, J.N. and Zellhart, J.W. (1995) Planning, in *Planning, Scheduling, and Control Integration in the Process Industries* (ed. C. Edward Bodington), McGraw-Hill, New York.

Gary, J.H. and Handwerk, G.E. (2001) *Petroleum Refining: Technology and Economics*, Marcel Dekker Inc., New York.

Grossmann, I.E. (2005) Enterprise-wide optimization: a new frontier in process systems engineering. *AIChE Journal*, **51**, 1846.

Grossmann, I.E., van den Heever, S.A., and Harjunkoski, I. (2001) Discrete optimization methods and their role in the integration of planning and scheduling. Proceedings of Chemical Process Control Conference 6 at Tucson, 2001, http://egon.cheme.cmu.edu/papers.html accessed on June 20, 2006.

Hartmann, J.C.M. (1998) Distinguish between scheduling and planning models. *Hydrocarbon Processing*, **77**, 5.

Himmelblau, D.M. (2008) Accounts of experiences in the application of artificial neural networks in chemical engineering. *Industrial & Engineering Chemistry Research*, **47**, 5782.

Jacob, S.M., Voltz, S.M., and Weekman, V. (1976) A lumping and reaction scheme for catalytic cracking. *AIChE Journal*, **22**, 701.

Jones, D.S.J. (1995) *Elements of Petroleum Processing*, John Wiley & Sons, New York.

Kay, J.W. and Titterington, D.M. (2000) *Statistics and Neural Networks*, Oxford University Press, Oxford, UK.

Lee, L.S., Chen, T.W., Haunh, T.N., and Pan, W.Y. (1989) Four lump kinetic model for fluid catalytic cracking process. *Canadian Journal of Chemical Engineering*, **67**, 615.

Maples, R.E. (1993) *Petroleum Refinery Process Economics*, Penn Well Publishing Company, Tulsa, Oklahoma.

Reklaitis, G.V. (1991) Perspectives on scheduling and planning of process operations. Proceedings of the fourth International symposium of Process System Engineering, Montebello, Canada.

Rippin, D.W.T. (1993) Batch process systems engineering: a retrospective and prospective review. *Computers & Chemical Engineering*, **17**, S1–S13.

Shah, N. (1998) Single- and multisite planning and scheduling: current status and future challenges. AIChE Symposium Series: Proceedings of the Third International Conference of the Foundations of Computer-Aided Process Operations, Snowbird, Utah, USA, July 5–10, 94, p. 75.

Venuto, P.B. and Habib, E.T. (1979) *Fluid Catalytic Cracking with Zeolite Catalysts*, Marcel Dekker Inc., New York.

Vermilion, W.L. (1971) Modern catalytic production of high octane gasoline and olefins. VOP Technical Seminar, Arlington Heights, Illinois.

Watson, K.M. and Nelson, E.F. (1933) Improved methods for approximating critical and thermal properties of petroleum fractions. *Industrial & Engineering Chemistry Research*, **25**, 880.

Weekman, V.W. and Nace, D.M. (1970) Kinetics of catalytic selectivity in fixed, moving and fluid bed reactors. *AIChE Journal*, **16**, 397.

Winston, W.L. and Venkataramanan, M. (2003) *Introduction to Mathematical Programming*, Brooks/Cole–Thomson Learning, California, USA.

Young, G.W., Swarez, W., Roberie, T.G., and Cheng, W.C. (1991) Reformulated gasoline: the role of current and future FCC catalysts. NPRA Annual Meeting, San Antonio.

3
Multisite Refinery Network Integration and Coordination

The use of mathematical programming models on an enterprise-wide scale to address strategic decisions considering various process integration alternatives yields substantial benefits. These benefits not only materialize in terms of economic considerations, but also in terms of process flexibility and improvements in the understanding of the process interactions and systems limitations.

In this chapter, we tackle the integration design and coordination of a multisite refinery network. The main feature of the chapter is the development of a simultaneous analysis strategy for process network integration through a mixed-integer linear program (MILP). The performance of the proposed model in this chapter is tested on several industrial-scale examples to illustrate the economic potential and trade-offs involved in the optimization of the network.

3.1
Introduction

With the current situation of high crude oil prices and the everlasting pressure to reduce prices of final petroleum products, refiners are faced with a very challenging situation. This nature of the petroleum economic environment provides a pressing motive for refineries to operate at an optimal level and continue to seek opportunities to increase their profit margin. This requires appropriate high level decision-making to utilize all available resources, not only on a single facility scale, but also in a more comprehensive outlook on an enterprise-wide scale. Such an approach provides an enhanced coordination and objectives alliance towards achieving a global optimal production strategy (Chopra and Meindl, 2004). The benefits projected from the coordination of multiple sites are not only in terms of expenses but also in terms of market effectiveness and responsiveness (Shah, 1998). Most of the time, there will be some necessity for a degree of independent management at each operating entity. However, the need for a coordinated response and the desire to minimize costs, imply that the various entities should be treated as parts of one large production system (Wilkinson, 1996). Planning for this system should be carried out centrally, allowing proper interactions between all operating facilities and, consequently, an

Planning and Integration of Refinery and Petrochemical Operations. Khalid Y. Al-Qahtani and Ali Elkamel
Copyright © 2010 WILEY-VCH Verlag GmbH & Co. KGaA, Weinheim
ISBN: 978-3-527-32694-5

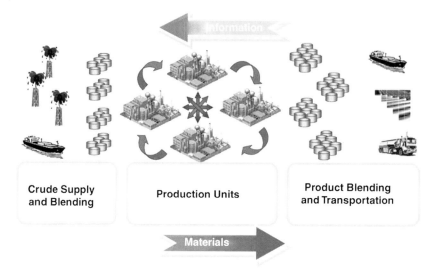

Figure 3.1 Refinery supply chain with process network integration.

efficient utilization of available resources. The understanding of such benefits has attracted much research in the areas of strategic planning in general and supply chain design and coordination in particular.

The objective of this chapter is to develop a methodology that can be applied for designing and analyzing process integration networks and production capacity expansions in a multiple refineries complex using different combinations of feedstocks, Figure 3.1. The integration strategy will allow the optimal coordination of the entire operating system through exchange of intermediate and product streams, as well as the efficient utilization of available resources in the different operating sites. The integration of utility streams between the different sites may be considered in a similar fashion. The proposed model is formulated as a mixed-integer linear programming (MILP) problem that minimizes the annualized operating and capital cost of the system. The application of the proposed methodology to achieve an integration and coordination strategy for the oil refining industry adds more complications and challenges. This is because refining is one of the most complex chemical industries comprising complicated processes and various configurations and structural alternatives. Although the MILP model is developed and applied to the refining industry in this book, it can be extended to any network of chemical processes. Since the decisions are of a long-term planning horizon, a linear model formulation is adequate to capture the required details of refinery processes (Zhang and Zhu, 2006). All capital cost investments are discounted over a time horizon in order to support a net present worth analysis.

Bok, Grossmann and Park (2000) explained a classification of chemical process networks and characterized them as either dedicated or flexible processes. Dedicated processes operate at one mode and for high volume products whereas flexible processes operate at different modes and produce different products at different times. In the formulation introduced in this chapter, we account for different operating modes. This should not be confused with flexible processes as the different modes in this study represent different product yields and do not require any set up costs or changeover times.

The remainder of this Chapter 3 is organized as follows. In Section 3.2 we will provide a literature review of process expansions and multisite planning and coordination studies in the chemical process industry. Then we will explain the problem statement and proposed model formulation in Sections 3.3 and 3.4, respectively. In Section 3.5, we will illustrate the performance of the model through various industrial-scale refinery examples and scenarios. The chapter ends with some concluding remarks in Section 3.6.

3.2
Literature Review

There has been quite a large stream of research concerned with capacity expansions and retrofit problems in the chemical and operations research literature. In this section, however, we will concentrate on expansion and strategic multisite planning studies. Single site short-term and mid-term planning and scheduling studies are beyond the scope of this book and the interested reader is referred to the work by Bodington and Baker (1990), Pinto, Joly and Moro (2000), and Kallrath (2005).

One of the early attempts in the operations research literature that considered multiple echelons and sites seems to be that of Williams (1981). Williams investigated different heuristic techniques of varying sophistication for production–distribution scheduling. Although he used simplifying assumptions, his work is one of the early attempts in the operations research literature to use coordinated planning across multiple echelons and sites. In the process systems engineering community, large scale multisite planning and coordination models have had a bigger share of the literature just recently.

Many earlier studies tackled different expansion problems in the chemical industry including the NLP formulation by Himmelblau and Bickel (1980) and the multiperiod MILP model by Grossmann and Santibanez (1980) and the recursive MILP model by Jimenez and Rudd (1987). A common drawback among these studies was their limitation of problem size due to computational burden. Other papers on capacity expansions can be found in Roberts (1964), Manne (1967) and Florian, Lenstra and Rinnooy Khan (1980) where discussions of how relevant these problems are to the industry are also provided. More recently, Sahinidis et al. (1989) developed a multiperiod MILP model for strategic planning in terms of selection and expansion of processes given forecasted demands and prices. In their study, they investigated several solution strategies to reduce model computational burden. The strategies included branch and bound, integer cuts, cutting planes, Benders decomposition and other heuristics. This model was later reformulated by Sahinidis and Grossmann (1991a, 1992) by identifying a lot sizing problem structure within the long-range problem formulation. The new reformulation improved the solution efficiency through tighter linear relaxation to the MILP model. The model was expanded to include continuous and batch flexible and dedicated processes (Sahinidis and Grossmann, 1991b). Along the same lines, Norton and Grossmann (1994) extended the work by Sahinidis *et al.* (1989) to account for dedicated and flexible processes in terms of feedstocks, products and the combination of continuous and batch processes. They illustrated their work with an example on a petrochemical complex.

Wilkinson, Shah and Pantelides (1996) proposed an approach that integrates production and distribution in multisite facilities using the resource task network representation proposed by Pantelides (1994). They applied this technique to an industrial case across a whole continent that involved production and distribution planning among different factories and markets. In a similar problem, McDonald and Karimi (1997) studied multiple semicontinuous facilities in a number of geographically distributed customers. The aim was to find an optimal allocation of recourses to tasks to meet a certain demand over a time horizon. They included a number of additional supply chain type constraints such as single sourcing, internal sourcing, and transportation times.

Iyer and Grossmann (1998) revisited the work by Sahinidis *et al.* (1989) and used a bilevel decomposition approach to reduce the computational complexity of the problem with the objective of solving larger scenarios. In a similar effort, Bok *et al.* (2000) extended the work by Norton and Grossmann (1994) to incorporate operational decisions, such as inventory profile, changeover cost, and intermittent supplies over multiple operating sites, over a short term horizon. They also used a bilevel decomposition approach to reduce computational time and illustrated their approach by several examples dealing with a petrochemical complex. Their model addressed short-term operating decisions and provided no insight on designing or retrofitting the process network.

Shah (1998) presented a review of the production planning and scheduling in single and multiple facilities. He pointed out that multisite problems have received little attention and are potential candidates for future research. Bunch, Rowe and Zentner (1998) developed the MILP model to find the lowest cost alternative among existing geographically distributed pharmaceutical facilities to satisfy a given demand. The model was used to find optimal assignment of products to facilities and production quantities over a time horizon. They used a commercial scheduling software (VirtECS) for both problem representation and solution. In their study, there was no clear underlying structure of the problem or systematic approach to the model formulation. Furthermore, the solution approach did not guarantee optimality.

Timpe and Kallrath (2000) developed a multi-period MILP model for a complete supply chain management of a multisite production network. The problem was formulated and applied to the food industry. The model concentrated on the coordination of the different echelons of the food supply chain and did not cover developing an integration scheme for the multisite facilities. Swaty (2002) studied the possibility of integrating a refinery and an ethylene plant through the exchange of process intermediate streams. The analysis was based on a linear programming (LP) model for each plant and profit marginal analysis of possible intermediate plant exchange. The study was implemented on a real life application in western Japan. Jackson and Grossmann (2003) proposed a multiperiod nonlinear programming model for production planning and distribution over multisite facilities. They used a temporal decomposition technique to reduce the scale of the problem.

Lasschuit and Thijssen (2003) pointed out the importance of developing integrated supply chain planning models on both strategic and tactical levels in the petroleum and chemical industry. They also stressed the issues that need to be accounted for when formulating these models. Neiro and Pinto (2004) proposed a general framework for modeling operational planning of the petroleum supply chain. They developed an MINLP model for the planning of multiple existing refineries, terminals and pipeline networks. Decisions included the selection of oil types and scheduling plan to the refineries subject to quality constraints as well as processing units, operating variables, product distribution, and inventory management. The model was applied to an industrial case in Brazil. Due to the high computational burden, the model was only solved for two time periods. The authors suggested that decomposition methods should be applied to yield a smaller MINLP or MILP model. Their approach did not consider the design problem of a network of operating facilities as their formulation addressed only operational type decisions.

Ryu and Pistikopoulos (2005) presented an MILP model for the design of enterprise-wide supply chains in the chemical industry. They investigated three different operating policies, namely, competition, cooperation and coordination. Their model was based on the assumption that plants will always have much larger capacities than demand. Furthermore, their work was mainly concerned with optimizing the operating policies among the different echelons of the supply chain and did not account for designing an integration policy between the multisite production facilities.

Khogeer (2005) developed an LP model for multiple refinery coordination. He developed different scenarios to experiment with the effect of catastrophic failure and different environmental regulation changes on the refineries performance. This work was developed using commercial planning software (Aspen PIMS). In his study, there was no model representation of the refineries systems or clear simultaneous representation of optimization objective functions. Such an approach deprives the study of its generalities and limits the scope to a narrow application. Furthermore, no process integration or capacity expansions were considered.

Another stream of research tackled modeling uncertainty in capacity expansion and supply chain studies in the process industry. Ierapetritou and Pistikopoulos (1994) developed a two-stage stochastic programming model for short to long-term planning problems. They proposed a decomposition method based on the generalized Benders decomposition algorithm where they used a Gaussian quadrature to estimate the expectation of the objective function. Liu and Sahinidis (1995, 1996, 1997) studied design uncertainty in process expansion through sensitivity analysis, stochastic programming and fuzzy programming, respectively. In their stochastic model, they used Monte Carlo sampling to calculate the expected objective function values. Their comparison over the different methodologies of including uncertainty was in favor of stochastic models when random parameter distributions are not available.

On a larger scale, Tsiakis, Shah and Pantelides (2001) developed a supply chain design for multiple markets and plants of steady-state continuous processes. Similarly, Ryu, Dua and Pistikopoulos (2004) presented a bilevel framework for planning an enterprise-wide network under uncertainty of some product demands on plant and warehouse capacities as well as resource availability. They considered the hierarchy of the supply chain and allowed for optimizing different levels of the chain individually.

The above discussion clearly points out the importance of multisite planning and indicates that such a problem is attracting a great deal of interest as the realization of the coordinated benefits became more vivid. However, to the best of our knowledge, no previous work has tackled developing a general framework for designing a network between multiple refineries in terms of material exchange. The aim of this study is to provide a methodology for the design of an integration and coordination policy of multiple refineries to explore potential synergies and efficient utilization of resources across an enterprise or multiple enterprises. The expansions of the facilities and the construction of the integration network are assumed to be implemented simultaneously in a single time horizon in order to minimize future process interruptions. Although our discussion will consistently refer to a network of refineries, the methodology that we will present can be readily extended to other chemical process networks.

3.3
Problem Statement

The optimization of refining processes involves a broad range of aspects varying from economical analysis and strategic expansions to crude oil selection, process levels targets, operating modes, and so on. The focus of this study is the development of a methodology for designing an integrated network and production expansion across multiple refineries as well as the establishment of an operating policy that sets feedstock combinations, process levels and operating mode preferences to satisfy a given demand. Such integration will provide appropriate means for improving the coordination across the whole network production system.

The general integration problem can be defined as:

A set of products $cfi \in CFR$ to be produced at multiple refinery sites $i \subset I$ is given. Each refinery consists of different production units $m \in M_{Ref}$ that can operate at different operating modes $p \in P$. An optimal feedstock from different available crudes $cr \in CR$ is desired. Furthermore, the process network across the multiple refineries is connected in a finite number of ways and an integration superstructure is defined. Market product prices, operating cost at each refinery, and product demands are assumed to be known.

The problem consists of determining the optimal integration for the overall network and associated coordination strategies across the refinery facilities as well as establishing an optimal overall production and determining the operating levels for each refinery site. The objective is to minimize the annualized cost over a given time horizon by improving the coordination and utilization of excess capacities in each facility. Expansion requirements to improve production flexibility and reliability are also considered.

For all refinery processes within the network we assume that all material balances are expressed in terms of linear yield vectors. Even though this might sound restrictive as most, if not all, refinery processes are inherently nonlinear, this practice is commonly applied in the petroleum refining business. Moreover, the decisions in this study are of a strategic level in which such linear formulation is adequate to address the required level of details involved at this stage. It is also assumed that processes have fixed capacities and the operating cost of each process and production mode is proportional to the process inlet flow. In the case of product blending, quality blending indices are used to maintain model linearity. Blending indices tables and graphs can often be found in petroleum refining books such as Gary and Handwerk (1994) or can be proprietorially developed by refining companies for their own use. It is also assumed that all products that are in excess of the local demand can be exported to a global market. Piping and pumping installation costs to transport intermediate streams as well as the operating costs of the new system were lumped into one fixed-charge cost. All costs are discounted over a 20 years time horizon and with an interest rate of 7%. No inventories will be considered since the model is addressing strategic decisions which usually cover a long period of time. We also assume perfect mixing and that the properties of each crude type are decided by specific key components. Properties of the oil mixture, such as viscosity, do not affect the strategic decisions of network design immensely and are therefore not considered in the model. Such considerations render the model unnecessarily complicated and are even tolerable on studies of operational planning and scheduling level (Lee et al., 1996; Jia and Ierapetritou, 2003, 2004).

3.4
Model Formulation

The model is formulated based on the state equipment network (SEN) representation (Yeomans and Grossmann, 1999). The general characterization of this representation includes three elements: state, task and equipment. A state includes all streams in a process and is characterized by either quantitative or qualitative attributes or both. The quantitative characteristics include flow rate, temperature and pressure, whereas the qualitative characteristics include other attributes such as the phase(s) of the streams. A task, on the other hand, represents the physical and chemical transformations that occur between consecutive states. Equipment provides the physical devices that execute a given task (e.g., reactor, absorber, heat exchanger).

The SEN allows for two types of task to equipment assignment: The first type is one task–one equipment (OTOE) assignment where tasks are assigned to equipment a priori. This type of assignment yields an identical representation to the state task network (STN) by Kondili, Pantelides and Sargent (1993). The second type is the variable task equipment assignment where the actual assignment of tasks to equipment is considered as part of the optimization problem. The use of this representation provides a consistent modeling strategy and an explicit handling of units that operate under different modes, which is common in the refining industry.

In treating stream mixing, the mixing device was defined as part of the designated refinery operation itself. This approach was also undertaken by Zhang and Zhu (2006). Therefore the only mixers considered are where the final blending takes place. This approach distinguishes the contribution of each feedstock to the final product. With this type of formulation, all variables and attributes of intermediate streams will depend on the crude type.

The problem is formulated as an MILP model where binary variables are used for designing the process integration network between the refineries and deciding on the production unit expansion alternatives. Linearity in the model was achieved by defining component flows instead of individual flows and associated fractions. The planning problem formulation is as follows.

3.4.1
Material Balance

The material streams, states, and their balances are divided into four categories; namely; raw materials, intermediates, products, and fuel system. All material balances are carried out on a mass basis. However, volumetric flow rates are used in the case where quality attributes of some streams only blend by volume.

Constraint (3.1) below illustrates the refinery raw materials balance in which throughput to each refinery crude distillation unit $p \in P'$ at plant $i \in I$ from each crude type $cr \in CR$ is equal to the available supply $S_{cr,i}^{Ref}$.

$$z_{cr,p,i} = S_{cr,i}^{Ref} \quad \forall\, cr \in CR, i \in I \quad \text{where} \tag{3.1}$$

$$p \in P' = \{\text{Set of CDU processes } \forall \text{ plant } i\}$$

The intermediate material balances within and across the refineries can be expressed as shown in constraint (3.2). The coefficient $\alpha_{cr,cir,i,p}$ can assume either a positive sign if it is an input to a unit or a negative sign if it is an output from a unit. The multirefinery integration matrix $\xi_{cr,cir,i,p,i'}$ accounts for all possible alternatives of connecting intermediate streams $cir \in CIR$ of crude $cr \in CR$ from refinery $i \in I$ to process $p \in P$ in plant $i' \in I'$. Variable $xi_{cr,cir,i,p,i'}^{Ref}$ represents the trans-shipment flow rate of crude $cr \in CR$, of intermediate $cir \in CIR$ from plant $i \in I$ to process $p \in P$ at plant $i' \in I$. The process network integration superstructure that constitutes all possible configuration structures can be defined a priori through suitable engineering and process analysis of all possible intermediate streams exchange.

$$\sum_{p \in P} \alpha_{cr,cir,i,p} z_{cr,p,i} + \sum_{i' \in I}\sum_{p \in P} \xi_{cr,cir,i',p,i}\, xi_{cr,cir,i',p,i}^{Ref}$$

$$- \sum_{i' \in I}\sum_{p \in P} \xi_{cr,cir,i,p,i'}\, xi_{cr,cir,i,p,i'}^{Ref} - \sum_{cfr \in CFR} w_{cr,cir,cfr,i} - \sum_{rf \in FUEL} w_{cr,cir,rf,i} = 0 \tag{3.2}$$

$$\forall\, cr \in CR, \quad cir \in CIR, \quad i\,\&\,i' \in I \quad \text{where} \quad i \neq i'$$

The material balance of final products in each refinery is expressed as the difference between flow rates from intermediate steams $w_{cr,cir,cfr,i}$ for each

$cir \in CIR$ that contribute to the final product pool and intermediate streams that contribute to the fuel system $w_{cr,cfr,rf,i}$ for each $rf \in FUEL$ as shown in constraint (3.3). In constraint (3.4) we convert the mass flow rate to volumetric flow rate by dividing it by the specific gravity $SG_{cr,cir}$ of each crude type $cr \in CR$ and intermediate stream $cir \in CB$. This is done as some quality attributes blend only by volume in the products blending pools.

$$\sum_{cr \in CR} \sum_{cir \in CB} w_{cr,cir,cfr,i} - \sum_{cr \in CR} \sum_{rf \in FUEL} w_{cr,cfr,rf,i} = x_{cfr,i}^{Ref} \quad \forall\, cfr \in CFR, \quad i \in I \tag{3.3}$$

$$\sum_{cr \in CR} \sum_{cir \in CB} \frac{w_{cr,cir,cfr,i}}{SG_{cr,cir}} = xv_{cfr,i}^{Ref} \quad \forall\, cfr \in CFR, \quad i \in I \tag{3.4}$$

Constraint (3.5) is the fuel system material balance where the term $cv_{rf,cir,i}$ represents the caloric value equivalent for each intermediate $cir \in CB$ used in the fuel system at plant $i \in I$. The fuel production system can either consist of a single or combination of intermediates $w_{cr,cir,rf,i}$ and products $w_{cr,cfr,rf,i}$. The matrix $\beta_{cr,rf,i,p}$ corresponds to the consumption of each processing unit $p \in P$ at plant $i \in I$ as a percentage of unit throughput.

$$\sum_{cir \in FUEL} cv_{rf,cir,i}\, w_{cr,cir,rf,i} + \sum_{cfr \in FUEL} w_{cr,cfr,rf,i}$$
$$- \sum_{p \in P} \beta_{cr,rf,i,p}\, z_{cr,p,i} = 0 \quad \forall\, cr \in CR, \quad rf \in FUEL, \quad i \in I \tag{3.5}$$

3.4.2
Product Quality

In general, the quality of a blend is composed of multiple components and is given by the following blending rule (Favennec et al., 2001):

$$Q = \frac{\sum q_i X_i}{\sum X_i}$$

where Q is the quality attribute of the blend, X_i is the quantity of each component in the blend, and q_i is the quality attribute of each blending component. However, when dealing with a large variety of blended products, as in the case of refining, we need to distinguish between attributes or components that blend by weight, such as sulfur content, and others that blend by volume, such as vapor pressure and octane number of gasoline. Furthermore, it is important to replace certain quality measurements such as viscosity values with certain blending indices in order to maintain model linearity. Blending indices tables and graphs can be found in petroleum refining books such as Gary and Handwerk (1994) or can be proprietorily developed by refining companies for their own use.

Constraints (3.6) and (3.7), respectively, express the lower and upper bound on quality constraints for all products that blend either by mass $q \in Q_w$ or by volume $q \in Q_v$.

$$\sum_{cr \in CR} \sum_{cir \in CB} \left(att_{cr,cir,q \in Qv} \frac{w_{cr,cir,cfr,i}}{sg_{cr,cir}} + att_{cr,cir,q \in Qw} \left[w_{cr,cir,cfr,i} - \sum_{rf \in FUEL} w_{cr,cfr,rf,i} \right] \right)$$
$$\geq q^L_{cfr,q \in Qv} xv^{Ref}_{cfr,i} + q^L_{cfr,q \in Qw} x^{Ref}_{cfr,i} \quad \forall \, cfr \in CFR, \quad q = \{Qw, Qv\}, \quad i \in I \tag{3.6}$$

$$\sum_{cr \in CR} \sum_{cir \in CB} \left(att_{cr,cir,q \in Qv} \frac{w_{cr,cir,cfr,i}}{sg_{cr,cir}} + att_{cr,cir,q \in Qw} \left[w_{cr,cir,cfr,i} - \sum_{rf \in FUEL} w_{cr,cfr,rf,i} \right] \right)$$
$$\leq q^U_{cfr,q \in Qv} xv^{Ref}_{cfr,i} + q^U_{cfr,q \in Qw} x^{Ref}_{cfr,i} \quad \forall \, cfr \in CFR, \quad q = \{Qw, Qv\}, \quad i \in I \tag{3.7}$$

3.4.3
Capacity Limitation and Expansion

Constraint (3.8) represents the maximum and minimum allowable flow rate to each processing unit. The coefficient $\gamma_{m,p}$ represents a zero-one matrix for the assignment of production unit $m \in M_{Ref}$ to process operating mode $p \in P$. As an example, the reformer is a production unit that can operate at high or low severity modes. The selection of the mode of operation will be considered as part of the optimization problem where variable task–equipment assignment (VTE) will be used. The term $AddC_{m,i,s}$ represents the additional expansion capacity for each production unit $m \in M_{Ref}$ at refinery $i \in I$ for a specific expansion size $s \in S$. Production systems expansion through the addition of new units requires detailed analysis and is usually quoted not only based on the unit flow rate but also on many other factors. For this reason, developing cost models of such expansions only as a function of the unit flow rate does not generally provide a good estimate. In our formulation, we only allowed the addition of predetermined capacities whose pricing can be acquired a priori through design companies' quotations. The integer variable $y \exp^{Ref}_{m,i,s}$ represents the decision to expand a production unit and it can take a value of one if the unit expansion is required or zero otherwise.

$$MinC_{m,i} \leq \sum_{p \in P} \gamma_{m,p} \sum_{cr \in CR} z_{cr,p,i} \leq MaxC_{m,i}$$
$$+ \sum_{s \in S} AddC_{m,i,s} \, y \exp^{Ref}_{m,i,s} \quad \forall \, m \in M_{Ref}, i \in I \tag{3.8}$$

Constraint (3.9) sets an upper bound on intermediate streams' flow rates between the different refineries. The integer variable $y \, pipe^{Ref}_{cir,i,i'}$ represents the decision of exchanging intermediate products between the refineries and takes on the value of one if the commodity is transferred from plant $i \in I$ to plant $i' \in I$ or zero otherwise, where $i \neq i'$. When an intermediate stream is selected to be exchanged between two refineries, its flow rate must be below the transferring pipeline capacity $F^U_{cir,i,i'}$.

$$\sum_{cr \in CR} \sum_{p \in P} \xi_{cr,cir,i,p,i'} \, xi^{Ref}_{cr,cir,i,p,i'} \leq F^U_{cir,i,i'} \, y\,pipe^{Ref}_{cir,i,i'} \quad (3.9)$$

$$\forall \, cir \in CIR, i' \, \& \, i \in I \quad \text{where} \quad i \neq i'$$

3.4.4
Product Demand

Constraint (3.10) stipulates that the final products from each refinery $x^{Ref}_{cfr,i}$ less the amount exported $e^{Ref}_{cfr',i}$ for each exportable product $cfr' \in PEX$ from each plant $i \in I$ must satisfy the domestic demand $D_{Ref_{cfr}}$.

$$\sum_{i \in I} \left(x^{Ref}_{cfr,i} - e^{Ref}_{cfr',i} \right) \geq D_{Ref_{cfr}} \quad \forall \, cfr \text{ and } cfr' \quad \text{where} \quad cfr \in CFR, cfr' \in PEX \quad (3.10)$$

3.4.5
Import Constraint

The imports or resources constraint (3.11) imposes upper and lower bounds on the available feedstock $cr \in CR$ to the refineries. The lower bound constraint might be useful in the cases where there are protocol agreements to exchange or supply crude oil between countries.

$$IM^L_{cr} \leq \sum_{i \in I} S^{Ref}_{cr,i} \leq IM^U_{cr} \quad \forall \, cr \in CR \quad (3.11)$$

3.4.6
Objective Function

The objective function considered in this study is given by:

$$\begin{aligned}
\text{Min} \quad & \sum_{cr \in CR} \sum_{i \in I} CrCost_{cr} \, S^{Ref}_{cr,i} \\
& + \sum_{p \in P} OpCost_p \sum_{cr \in CR} \sum_{i \in I} z_{cr,p,i} \\
& + \sum_{cir \in CIR} \sum_{i \in I} \sum_{i' \in I} InCost_{i,i'} \, y\,pipe^{Ref}_{cir,i,i'} \quad \text{where } i \neq i' \\
& + \sum_{i \in I} \sum_{m \in M_{Ref}} \sum_{s \in S} InCost_{m,s} \, y\,exp^{Ref}_{m,i,s} \\
& - \sum_{cfr \in PEX} \sum_{i \in I} Pr^{Ref}_{cfr} \, e^{Ref}_{cfr,i}
\end{aligned} \quad (3.12)$$

The above objective represents a minimization of the annualized cost which comprises crude oil cost, refineries operating cost, refineries intermediate exchange piping cost, production system expansion cost, and export revenue. The operating cost of each process is assumed to be proportional to the process inlet flow and is expressed on a yearly basis.

3.5
Illustrative Case Study

In this section, we present two examples with different scenarios. The first example illustrates the performance of the model on a single site total refinery planning problem where we compare the results of the model to an industrial scale study from Favennec et al. (2001). This example serves to validate our model and to make any necessary adjustments. The second example extends the scale of the model application to cover three complex refineries in which we demonstrate the different aspects of the model. The refineries considered are of large industrial-scale refineries and actually mimic a general set-up of many areas around the world. The decisions in this example include the selection of crude blend combination, design of process integration network between the three refineries, and decisions on production units' expansion options and operating levels.

The modeling system GAMS (Brooke et al., 1996) was used for setting up the optimization models and the problems are solved by BDMLP 1.3 on a Pentium M processor 2.13 GHz.

3.5.1
Single Refinery Planning

Figure 3.2 provides a SEN representation of the refinery considered in this example. The planning horizon was set to one month in order to compare the model results with those of Favennec et al. (2001). As shown in Figure 3.2, the refinery uses two different feedstocks (e.g., Arabian light and Kuwait crudes) where the optimum blend is used to feed the atmospheric crude unit. The atmospheric crude unit separates crude oil into several fractions including LPG, naphtha, kerosene, gas oil and residues. The heavy residues are then sent to the vacuum unit where they are further separated into vacuum gas oil and vacuum residues. Depending on the production targets, different processing and treatment processes are applied to the crude fractions. In this example, the naphtha is further separated into heavy and light naphtha. Heavy naphtha is sent to the catalytic reformer unit to produce high octane reformates for gasoline blending and light naphtha is sent to the light naphtha pool and to an isomerization unit to produce isomerate for gasoline blending. The middle distillates are combined with other similar intermediate streams and sent for hydrotreating and then for blending to produce jet fuels and gas oils. Atmospheric and vacuum gas oils are further treated by catalytic cracking and in other cases by hydrocracking or both to increase the gasoline and distillate

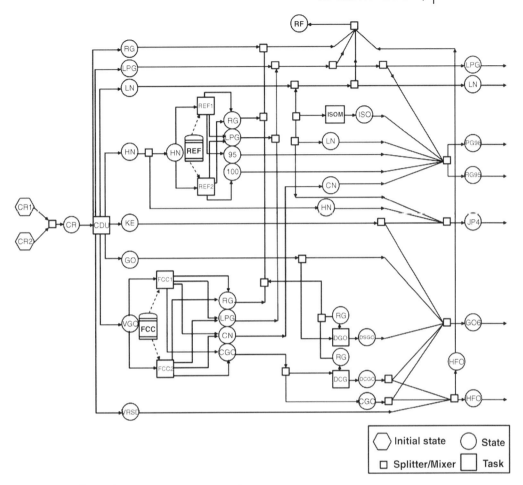

Figure 3.2 Refinery 1 layout using SEN representation.

yields. In some refineries, vacuum residues are further treated using cooking and thermal processes to increase light products yields. The final products in this example consist of liquefied petroleum gas (LPG), light naphtha (LT), two grades of gasoline (PG98 and PG95), No.4 jet fuel (JP4), No.6 gas oil (GO6), and heating fuel oil (HFO). The major capacity limitations as well as availability constraints are shown in Table 3.1.

We slightly adjusted our model to allow for spot market buying and selling of heavy naphtha, vacuum gas oil and all products in order to demonstrate actual total site refinery planning and compare our results with those of Favennec et al. (2001). The model results and a comparison are shown in Table 3.2.

Table 3.1 Major refinery capacity constraints for single refinery planning.

	Lower limit 1000 t/month	Higher limit 1000 t/month
Production Capacity		
Distillation	—	700
Reformer		
95 severity	2	—
Total	—	60
Isomerization	—	15
Fluid catalytic cracker	—	135
Total Desulfurization	—	150
Crude availability		
Crude 1	—	400
Crude 2	260	—

This example illustrates the capability and flexibility of our formulation to capture the details of a tactical or a medium-term planning horizon of one month. There are some minor differences in the results. This is because we were not able to access the detailed model used in their study and hence were not able to match their assumptions.

However, our proposed modeling methodology is of great benefit to refiners as they can align their long and medium plans through a common purpose model.

Table 3.2 Model results and comparison of single refinery planning.

Process variables		Results (1000 t/month)	
		Case Study	Proposed model
Crude oil supply	Crude 1	278.6	268.1
	Crude 2	260	260
	Total	538.6	528.1
Production levels	Crude	538.6	528.1
	Reformer 95	2	2
	Reformer 100	57.72	58.00
	Isomerization	11.72	11.63
	FCC gasoline mode	0	0
	FCC gas oil mode	130.5	128.4
	Des Gas oil	119.9	118.2
	Des cycle gas oil	23.8	23.0
Intermediate import	Heavy naphtha	7.38	8.62
Final product import	GO6	0	1
Exports	PG95	12.78	13.6
	JP4	5	0
Total cost ($/month)		90 177	91 970

3.5.2
Multisite Refinery Planning

In this example, we extend the scale of the case study to cover strategic planning for three complex refineries by which we demonstrate the performance of our model under different considerations. The three refineries considered represent industrial-scale size refineries and an actual configuration that can be found in many industrial sites around the world. See Figures 3.3 and 3.4 for the second and third refinery layouts, respectively. These are in addition to the refinery case study of the single refinery planning in Section 3.5.1. The three refineries are assumed to be in one industrial area, which is a common situation in many locations around the world. The refineries are coordinated through a main headquarters, centralized planning is assumed, and feedstock supply is shared. The final products of the three

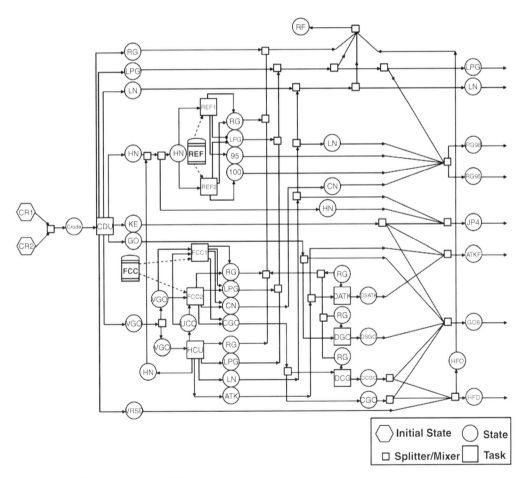

Figure 3.3 Refinery 2 layout using SEN representation.

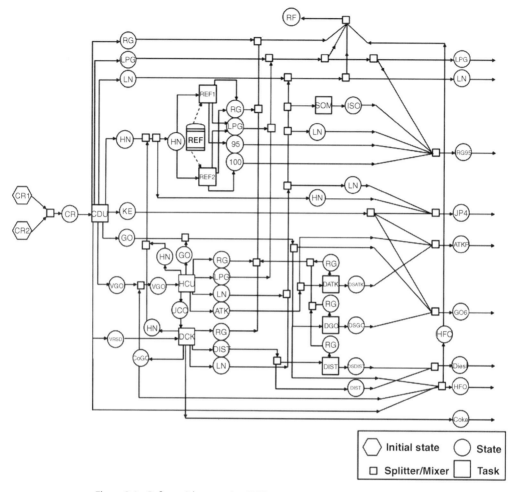

Figure 3.4 Refinery 3 layout using SEN representation.

refineries consist of liquefied petroleum gas (LPG), light naphtha (LT), two grades of gasoline (PG98 and PG95), No. 4 jet fuel (JP4), military jet fuel (ATKP), No.6 gas oil (GO6), diesel fuel (Diesel), heating fuel oil (HFO), and petroleum coke (coke). We will now consider several practical scenarios to demonstrate the advantage of the proposed integration model and its robustness under different considerations.

3.5.2.1 Scenario-1: Single Feedstock, Multiple Refineries with No Integration

In this scenario, the three refineries are using a single feedstock type, Arabian Light, and operate centrally with no network integration alternatives. The major model constraints and results are shown in Tables 3.3 and 3.4, respectively. The three refineries collaborate to satisfy a given local market demand where the

Table 3.3 Major refineries capacity constraints for multisite refinery planning, Scenario-1 and 2.

Production Capacity	Higher limit (1000 t/year)		
	R1	R2	R3
Distillation	4500	12 000	9900
Reforming	1000	2000	1800
Isomerization	200	—	450
Fluid catalytic cracker	1700	1700	—
Hydrocracker	—	2000	2500
Delayed coker	—	—	2200
Des gas oil	1900	3000	2400
Des cycle gas oil	200	750	—
Des ATK	—	1200	1680
Des Distillates	—	—	700
Crude availability			
Crude 1	31 200		
Crude 2	—		
Local Demand			
LPG	432		
LN	312		
PG98	540		
PG95	4440		
JP4	2340		
GO6	4920		
ATK	1800		
HFO	200		
Diesel	480		
Coke	300		

model provides the production and blending level targets for the individual sites. Products that exceeded local market demand are either sold in the spot market or exported. The annual production cost across the facilities was found to be $7 118 000.

3.5.2.2 Scenario-2: Single Feedstock, Multiple Refineries with Integration

In this scenario, we allowed the design of an integration network between the three refineries using the same set of constraints as in Scenario-1. The cost parameters for pipelines installation were calculated as cost per distance between the refineries, and then multiplied by the required pipe length in order to connect any two refineries. The pipeline diameter considered in all cases was 8 inches.

As shown from the results with this scenario in Table 3.5, by allowing the design of an optimal integration network between the refineries, we were able to achieve annual savings exceeding $230 000. The savings will increase as the number of plants, production units and integration alternatives across the enterprise increase. However, benefits are not only limited to reducing cost, but also include improved

Table 3.4 Model results of multisite refinery planning; Scenario-1.

Process variables		Results (1000 t/year)		
		R1	R2	R3
Crude oil supply	Crude 1	4500	12 000	9900
	Crude 2	—	—	—
	Total	4500	12 000	9900
Production levels	Crude unit	4500	12 000	9900
	Reformer 95	163.0	250.0	502.2
	Reformer 100	410.1	1574.6	1239.5
	Isomerization	140.3	—	450
	FCC gasoline mode	954	899.3	—
	FCC gas oil mode	—	—	—
	Hydrocracker	—	1740.4	2098.8
	Delayed coker	—	—	1402.4
	Des Gas oil	1395	2753.9	2383.8
	Des cycle gas oil	200	401.1	—
	Des ATK	—	1200	1447.1
	Des Distillates	—	—	338
Exports	PG95	214.3		
	JP4	1427		
	GO6	3540.1		
	HFO	1917.4		
	ATK	1508		
	Coke	176.8		
	Diesel	2.40		
Total cost ($/yr)		7 118 000		

flexibility and sustainability of production as well as proper utilization and allocation of resources among the refineries network. As an example, diesel production in the first scenario (no integration) was barely satisfying local demand of 480 000 t/year and only 2400 t/year were left for export. With such a thin production margin, the plant did not have enough flexibility to face variations in diesel demand. In Scenario-2, however, the production margin of diesel increased from 2400 to 320 600 t/year. The benefits were in terms of increasing exports and, hence, profit, and also gaining more diesel production flexibility to meet any variations in local market demand.

3.5.2.3 Scenario-3: Multiple Feedstocks, Multiple Refineries with Integration

In this scenario, we provide multiple feedstocks, namely; Arabian Light and Kuwait crude, to the refineries complex and demonstrate the selection of crude combinations to each refinery as well as how the integration network will change. All sets of constraints remained the same except for the crude supply as we imposed a higher availability limit of 20 000 t/year of each crude type. The results of the model are shown in Table 3.6.

3.5 Illustrative Case Study | 73

Table 3.5 Model results of multisite refinery planning; Scenario-2.

Process variables			Results (1000 t/year)		
			R1	R2	R3
Crude oil supply	Crude 1		4500	12 000	9900
	Crude 2		—	—	—
	Total		4500	12 000	9900
Production levels	Crude unit		4500	12 000	9900
	Reformer 95		—	—	887.2
	Reformer 100		573.1	2000	686.9
	Isomerization		140.3	—	450
	FCC gasoline mode		616.3	1500	—
	FCC gas oil mode		—	—	—
	Hydrocracker		—	1105	2436.54
	Delayed coker		—	—	2066
	Des Gas oil		1390	2822.6	2383.8
	Des cycle gas oil		200	669	—
	Des ATK		—	762	1680
	Des Distillates		—	—	498
Intermediate streams exchange	From	R1	VGO	—	—	337.7
			VRSD	—	—	350
		R2	VRSD	—	—	319.5
		R3	HN	—	283.6	—
			CoGO	—	350	—
Exports	PG95		334.1		
	JP4		1433.6		
	GO6		3862		
	HFO		1203		
	ATK		1251.7		
	Coke		402.4		
	Diesel		230.6		
Total cost ($/yr)			6 885 000		

The total crude oil supply to the refineries complex remained the same. However, the overall utilization of some major production units has changed. The capacity utilization of thermal and catalytic reactors has increased whereas reforming utilization has slightly decreased. This is because Kuwait crude contains more heavy ends than Arabian light which has more of the lighter ends. The selection of crude supply type was in favor of Arabian Light as it was processed up to its maximum availability level of 20 000 t/year where the remaining required crude was fulfilled by Kuwait crude. Due to the shortage in Arabian Light supply, the model used Kuwait crude to satisfy local market demand although it yields a higher overall annual production cost of $ 7 263 000. This scenario illustrates the use of different combinations of crude types and how this affects the overall utilization of production units, refineries network integration, and total annual cost. In the next scenario we will see how expanding some processes will increase the utilization of other production units' capacity.

Table 3.6 Model results of multisite refinery planning; Scenario-3.

Process variables			Results (1000 t/year)		
			R1	R2	R3
Crude oil supply	Crude 1		3536	11 835	4629
	Crude 2		964	—	5271
	Total		4500	11 835	9900
Production levels	Crude unit		4500	11 835	9900
	Reformer 95		227	—	—
	Reformer 100		736	1804	1074
	Isomerization		144	—	450
	FCC gasoline mode		485	1395	—
	FCC gas oil mode		—	—	—
	Hydrocracker		—	1741	2431
	Delayed coker		—	—	2066
	Des Gas oil		1121	2727	2127
	Des cycle gas oil		200	622	—
	Des ATK		—	1200	1676
	Des Distillates		—	—	498
Intermediate streams exchange	From	R1	VGO	—	265	—
			VRSD	—	—	24
		R3	HN	226	—	—
Exports	JP4			837	
	GO6			2860	
	HFO			2390	
	ATK			1794	
	Coke			402.4	
	Diesel			230.6	
Total cost ($/yr)				7 263 000	

3.5.2.4 Scenario-4: Multiple Feedstocks, Multiple Refineries with Integration and Increased Market Demand

In all previous scenarios, we did not change the market demand and, therefore, there was no expansion in the production unit capacities of the refineries. In this scenario, we will simulate a change in market demand and examine the modifications suggested by the model. Table 3.7 illustrates the new major operating constraints.

In general, developing cost models of production system expansions only as a function of the unit flow rate does not provide a good capital cost estimate. For this reason, and as we mentioned earlier, our formulation only allowed the addition of predetermined capacities whose price can be acquired a priori through design companies' quotations.

Table 3.8 shows the new strategic plan for all refineries in terms of crude oil supply combinations, production expansions, and integration network design between the refineries. In response to the increase in the diesel production requirements by more

Table 3.7 Major refineries capacity constraints for multisite refinery planning, Scenario-4.

Production Capacity	Higher limit (1000 t/year)		
	R1	R2	R3
Distillation	5000	12 000	11 000
Reforming	1000	2000	1800
Isomerization	200	—	450
Fluid catalytic cracker	1700	1700	—
Hydrocracker	—	2000	2500
Delayed coker	—	—	2200
Des gas oil	1900	3000	2400
Des cycle gas oil	200	750	—
Des ATK	—	1200	1680
Des Distillates	—	—	700
Crude availability			
Crude 1	20 000		
Crude 2	20 000		
Local Demand			
LPG	432		
LN	312		
PG98	540		
PG95	4440		
JP4	2340		
GO6	4920		
ATK	1800		
HFO	200		
Diesel	1200		
Coke	300		

than two-fold, the new plan suggests the installation of a new thermal coker and a distillates desulfurization unit in Refinery 3. This change has a clear effect on the integration network design among the refineries. As an illustration, the new plan suggests to increase the level of intermediate exchange of vacuum residues from Refineries 1 and 2 to Refinery 3 in order to efficiently utilize the additional capacities of the coker and distillates desulfurization units. The total annual cost has increased to $ 21 463 000 due to the capital and operating costs of the additional units.

3.6
Conclusion

A mixed-integer programming model for minimizing cost in the strategic planning of a multiple refineries network was presented. The objective was to develop a methodology for designing a process integration network and production capacity expansions in a multiple refinery complex using different feedstock alternatives. Two

Table 3.8 Model results of multisite refinery planning; Scenario-4.

Process variables				Results (1000 t/year)		
				R1	R2	R3
Crude oil supply	Crude 1			3077	8293	6059
	Crude 2			1923	1333	4420
	Total			5000	9626	10 479
Production levels	Crude unit			5000	9626	10 479
	Reformer 95			276	—	19
	Reformer 100			724	2000	610
	Isomerization			163	—	450
	FCC gasoline mode			751	1595	—
	FCC gas oil mode			—	—	—
	Hydrocracker			—	1065	2500
	Delayed coker			—	—	3490
	Des Gas oil			1900	3000	1089
	Des cycle gas oil			200	712	—
	Des ATK			—	1200	1258
	Des Distillates			—	—	841
Intermediate streams exchange	From	R1	GO	—	52	—
			VGO	—	216	—
			VRSD	—	—	400
		R2	VRSD	—	—	400
		R3	HN	242	332	—
			GO	400	400	—
			ATK	—	240	—
			UCO	—	23	—
Process Expansions	Delayed Coker	—	—	1380		
	Des Distillates	—	—	600		
Exports	JP4			543		
	GO6			2655		
	HFO			1217		
	ATK			1272		
	Coke			886		
	Diesel			—		
Total cost ($/yr)				21 463 000		

examples with multiple scenarios of large-scale refineries were solved to illustrate the performance of the proposed design methodology and to show the economic potential and trade-offs involved in the optimization of such systems. The integration specifically addressed intermediate material transfer between processing units at each site. In the formulation, bilinear mixing equations were avoided by introducing individual component flows in order to maintain linearity.

Petroleum refining is a central and crucial link in the oil supply chain and has received extensive attention over the last decades. However, despite all the progress that has been made in developing planning and scheduling models, a general purpose model is still a target (Grossmann, 2005). A general model that can be used for different planning levels, short, medium and long range will be of great benefit in terms of seamless interactions of these functions. In this work we showed the capability of the proposed model in capturing all details required for medium-range and tactical planning, as illustrated by Example 1. This is a step forward in achieving such vertical integration among all planning hierarchies.

In this chapter, all parameters were assumed to be deterministic. However, the current situation of fluctuating petroleum crude oil prices and demands is an indication that markets and industries everywhere are impacted by uncertainties. For example, source and availability of crude oils as the raw material; prices of feedstock, chemicals, and commodities; production costs; and future market demand for finished products will have a direct impact on final decisions. Thus, acknowledging the shortcomings of deterministic models, the next Chapters will consider uncertainties in the design problem.

References

Bodington, C.E. and Baker, T.E. (1990) A history of mathematical programming in the petroleum industry. *Interfaces*, **20**, 117.

Bok, J.K., Grossmann, I.E., and Park, S. (2000) Supply chain optimization in continuous flexible processes. *Industrial & Engineering Chemistry Research*, **39**, 1279.

Brooke, A., Kendrick, D., Meeraus, A., and Raman, R. (1996) *GAMS–A User's Guide*, GAMS Development Corporation, Washington DC.

Bunch, P.R., Rowe, R.L., and Zentner, M.G. (1998) Large scale multi-facility planning using mathematical programming methods. AIChE Symposium Series, *Proceedings of the Third International Conference of the Foundations of Computer-Aided Process Operations. Snowbird, Utah, USA, July 5–10*, American Institute of Chemical Engineering, **94**, p. 249.

Chopra, S. and Meindl, P. (2004) *Supply Chain Management: Strategy, Planning, and Operations*, 2nd edn, Pearson Education, New Jersey.

Favennec, J., Coiffard, J., Babusiaux, D., and Trescazes, C. (2001) *Refinery Operation and Management*, vol. **6**, Editions Technip, Paris.

Florian, M.K., Lenstra, J.K., and Rinnooy Khan, A.H. (1980) Deterministic production planning: Algorithms and complexity. *Management Science*, **26**, 669.

Gary, J.H. and Handwerk, G.E. (1994) *Petroleum Refining: Technology and Economics*, Marcel Dekker Inc., New York.

Grossmann, I.E. (2005) Enterprise-wide optimization: a new frontier in process systems engineering. *AIChE Journal*, **51**, 1846.

Grossmann, I.E. and Santibanez, J. (1980) Application of mixed-integer linear programming in process synthesis. *Computers & Chemical Engineering*, **4**, 205.

Himmelblau, D.M. and Bickel, T.C. (1980) Optimal expansion of a hydrodesulfurization process. *Computers & Chemical Engineering*, **4**, 101–112.

Ierapetritou, M.G. and Pistikopoulos, E.N. (1994) Novel optimization approach of stochastic planning models. *Industrial & Engineering Chemistry Research*, **33**, 1930.

Iyer, R.R. and Grossmann, I.E. (1998) A bilevel decomposition algorithm for long-range planning of process networks. *Industrial & Engineering Chemistry Research*, **37**, 474.

Jackson, J.R. and Grossmann, I.E. (2003) Temporal decomposition scheme for nonlinear multisite production planning and distribution models. *Industrial & Engineering Chemistry Research*, **42**, 3045.

Jia, Z. and Ierapetritou, M. (2003) Refinery short-term scheduling using continuous time formulation: crude-oil operations.

Industrial & Engineering Chemistry Research, 42, 3085.

Jia, Z. and Ierapetritou, M. (2004) Efficient short-term scheduling of refinery operations based on a continuous time formulation. Computers & Chemical Engineering, 28, 1001.

Jimenez, G.A. and Rudd, D.F. (1987) Use of a recursive mixed-integer programming model to detect an optimal integration sequence for the Mexican petrochemical industry. Computers & Chemical Engineering, 11, 291.

Kallrath, J. (2005) Solving planning and design problems in the process industry using mixed integer and global optimization. Annals of Operations Research, 140, 339.

Khogeer, A.S. (2005) Multiobjective multirefinery optimization with environmental and catastrophic failure effects objectives. Ph.D. Thesis. Colorado State University.

Kondili, E., Pantelides, C.C., and Sargent, R.W.H. (1993) A general algorithm for short-term scheduling of batch operations-I. MILP formulation. Computers & Chemical Engineering, 17, 211.

Lasschuit, W. and Thijssen, N. (2003) Supporting supply chain planning and scheduling decisions in the oil & chemical industry, in Proceedings of Fourth International Conference on Foundations of Computer-Aided Process Operations (eds I.E. Grossmann and C.M. McDonald), Coral Springs, CAChE, p. 37.

Lee, H., Pinto, J.M., Grossmann, I.E., and Park, S. (1996) Mixed-integer linear programming model for refinery short-term scheduling of crude oil unloading with inventory management. Industrial & Engineering Chemistry Research, 35, 1630.

Liu, M.L. and Sahinidis, N.V. (1995) Computational trends and effects of approximations in an MILP model for process planning. Industrial & Engineering Chemistry Research, 34, 1662.

Liu, M.L. and Sahinidis, N.V. (1996) Optimization in process planning under uncertainty. Industrial & Engineering Chemistry Research, 35, 4154.

Liu, M.L. and Sahinidis, N.V. (1997) Process planning in a fuzzy environment. European Journal of Operational Research, 100, 142.

Manne, A.S. (1967) Investment for Capacity Expansion Problems, MIT Press, Cambridge, MA.

McDonald, C.M. and Karimi, I.A. (1997) Planning and scheduling of parallel semicontinuous processes. 1. Production planning. Industrial & Engineering Chemistry Research, 36, 2691.

Neiro, S.M.S. and Pinto, J.M. (2004) A general modeling framework for the operational planning of petroleum supply chains. Computers & Chemical Engineering, 28, 871.

Norton, L.C. and Grossmann, I.E. (1994) Strategic planning model for complete process flexibility. Industrial & Engineering Chemistry Research, 33, 69.

Pantelides, C.C. (1994) Unified frameworks for optimal process planning and scheduling, in Proceeding of the Second Conference on Foundations of Computer-Aided Operations, CAChE Publications, p. 253.

Pinto, J.M., Joly, M., and Moro, L.F.L. (2000) Planning and scheduling models for refinery operations. Computers & Chemical Engineering, 24, 2259.

Roberts, S.M. (1964) Dynamic Programming in Chemical Engineering and Process Control, Academic Press, New York.

Ryu, J. and Pistikopoulos, E.N. (2005) Design and operation of an enterprise-wide process network using operation policies. 1. Design. Industrial & Engineering Chemistry Research, 44, 2174.

Ryu, J., Dua, V., and Pistikopoulos, E.N. (2004) A bilevel programming framework for enterprise-wide process networks under uncertainty. Computers & Chemical Engineering, 28, 1121.

Sahinidis, N.V. and Grossmann, I.E. (1991a) Reformulation of multiperiod MILP models for planning and scheduling of chemical processes. Computers & Chemical Engineering, 15, 255.

Sahinidis, N.V. and Grossmann, I.E. (1991b) Multiperiod investment model for processing networks with dedicated and flexible plants. Industrial & Engineering Chemistry Research, 30, 1165.

Sahinidis, N.V. and Grossmann, I.E. (1992) Reformulation of multi-period MILP model for planning and scheduling of chemical processes. Operations Research, 40, S127.

Sahinidis, N.V., Grossmann, I.E., Fornari, R.E., and Chathrathi, M. (1989) Optimization model for long range planning in chemical industry. *Computers & Chemical Engineering*, **13**, 1049.

Shah, N. (1998) Single- and multisite planning and scheduling: current status and future challenges. AIChE Symposium Series: *Proceedings of the Third International Conference of the Foundations of Computer-Aided Process Operations, Snowbird, Utah, USA, July 5–10*, American Institute of Chemical Engineering, **94**, p. 75.

Swaty, T.E. (2002) Consider over-the-fence product stream swapping to raise profitability. *Hydrocarbon Processing*, **81**, 37.

Timpe, C.H. and Kallrath, J. (2000) Optimal planning in large multisite production networks. *European Journal of Operational Research*, **126**, 422–435.

Tsiakis, P., Shah, N., and Pantelides, C.C. (2001) Design of multiechelon supply chain networks under demand uncertainty. *Industrial & Engineering Chemistry Research*, **40**, 3585.

Wilkinson, S.T. (1996) Aggregate Formulation for Large-Scale Processing Scheduling Problems, Ph.D. Thesis. Imperial College.

Wilkinson, S.J., Shah, N., and Pantelides, C.C. (1996) Integrated production and Distribution shedueling on a Europe-Wide Basis. *Computers & Chemical Engineering*, **S20**, S1275.

Williams, J.F. (1981) Heuristic techniques for simultaneous scheduling of production and distribution in multi-echelon structures. *Management Science*, **27**, 336.

Yeomans, H. and Grossmann, I.E. (1999) A systematic modeling framework of superstructure optimization in process synthesis. *Computers & Chemical Engineering*, **23**, 709.

Zhang, N. and Zhu, X.X. (2006) Novel modeling and decomposition strategy for total site optimization. *Computers & Chemical Engineering*, **30**, 765.

4
Petrochemical Network Planning

The structure of the petrochemical industry is a highly interactive and complex structure as it involves hundreds of chemicals and processes with products of one process being the feedstocks of many others. For most chemicals, the production route from feedstock to final products is not unique, but includes many possible alternatives. As complicated as it may seem, this structure is comprehensible, at least in a general sense.

This chapter explains the general representation of a petrochemical planning model which selects the optimal network from the overall petrochemical superstructure. The system is modeled as a mixed-integer linear programming (MILP) problem and illustrated via a numerical example.

4.1
Introduction

The petrochemical industry is a network of highly integrated production processes where products of one plant may have an end use or may also represent raw materials for other processes. This flexibility in petrochemical products production and the availability of many process technologies offer the possibility of switching between production methods and raw materials utilization. The world economic growth and increasing population will keep global demand for transportation fuels and petrochemical products growing rapidly for the foreseeable future. One half of the petroleum consumption over the period 2003–2030 will be in the transportation sector, whereas the industrial sector accounts for 39% of the projected increase in world oil consumption, mostly for chemical and petrochemical processes (EIA, 2006). Meeting this demand will require large investments and proper strategic planning for the petrochemical industry.

The objective of this chapter is to give an overview of the optimization of petrochemical networks and to set up the deterministic model which will be used in the analysis of parameter uncertainties in Chapter 8.

The remainder of this Chapter 4 is organized as follows. In Section 4.2 we will review the related literature in the area of petrochemical planning. Then, we will

Planning and Integration of Refinery and Petrochemical Operations. Khalid Y. Al-Qahtani and Ali Elkamel
Copyright © 2010 WILEY-VCH Verlag GmbH & Co. KGaA, Weinheim
ISBN: 978-3-527-32694-5

discuss the deterministic model formulation for the petrochemical network planning and illustrate its performance through an industrial case study in Sections 4.3 and 4.4, respectively. The chapter ends with concluding remarks in Section 4.5.

4.2
Literature Review

The realization of the need for petrochemical planning along with its important impact has inspired a great deal of research in order to devise different modeling frameworks and algorithms. These include optimization models with continuous and mixed-integer programming under both deterministic and uncertainty considerations.

The seminal work of Stadtherr and Rudd (1976, 1978) defined the petrochemical industry as a network of chemical process systems with linear chemical transformations and material interactions. They showed that the model provided a good representation of the petrochemical industry and can be used as a tool for estimating the relative effectiveness of available and new technologies and their impact on the overall industry. Their objective was to minimize feedstock consumption. A similar LP modeling approach was adapted by Sokic and Stevancevic (1983). Sophos, Rotstein and Stephanopoulos (1980) presented a model that minimizes feedstock consumption and entropy creation (lost work). Fathi-Afshar and Yang (1985) devised a multiobjective model of minimizing cost and gross toxicity emissions. Modeling the petrochemical industry using linear programming may have shown its ability to provide relatively reliable results through different technology structures. However, the need for approximating non-linear objective functions or the restriction of process technology combination alternatives mandated different modeling techniques involving mixed-integer programming.

Some of the first mixed-integer programming models that tackled this problem were proposed by Jimenez, Rudd and Meyer (1982) and Jimenez and Rudd (1987) for the development of the Mexican petrochemical industry. The proposed models were used to plan the installation of new plants with profitable levels as opposed to importing chemical products. However, there were no capacity limitation constraints on the processes. Al-Amir, Al-Fares and Rahman (1998) developed an MILP model for the development of Saudi Arabia's petrochemical industry maximizing profit. The model included the minimum economic production quantity for the different processes and accounted for domestic consumption and global market exports. This model was further extended by Al-Fares and Al-Amir (2002) to include four main product categories: propylene, ethylene, synthesis gas and aromatics and their derivatives. They devised a non-linear objective function of production investment cost at different production levels and derived a linear representation of the function through piece-wise linear approximation. Al-Sharrah *et al.* (2001), and Al-Sharrah, Alatiqi and Elkamel (2002) presented MILP models that took sustainability and strategic technology selection into consideration. The models included a constraint to limit the selection of one technology to produce a chemical achieving a long term financial stability and an environmental consideration through a suitable objective.

Sustainability was quantified by a health index of the chemicals and increasing profit was represented by the added value of each process in the network. This work was later extended by Al-Sharrah, Alatiqi and Elkamel (2003) with the aim of identifying long-range and short-range disturbances that affect planning in the petrochemical industry. Al-Sharrah, Hankinson and Elkamel (2006) further developed their petrochemical planning framework into a multiobjective model accounting for economic gain and risk from plant accidents. The above body of research did not take into account parameter uncertainties.

The above discussion shows the importance of petrochemical network planning in process system engineering studies. In this chapter we develop a deterministic strategic planning model of a network of petrochemical processes. The problem is formulated as a mixed-integer linear programming model with the objective of maximizing the added value of the overall petrochemical network.

4.3
Model Formulation

The optimization of petrochemical network design involves a broad range of aspects, including economic and environmental analysis, strategic selection of processes and production capacities. The deterministic model presented in this study is slightly modified from that of Al-Sharrah *et al.* (2001), Al-Sharrah, Hankinson and Elkamel (2006). A set of CP number of chemicals involved in the operation of M_{pet} processes is assumed to be given. Let x_m^{Pet} be the annual level of production of process $m \in M_{pet}$, F_{cp}^{Pet} the amount of chemical $cp \in CP$ as a feedstock, and $\delta_{cp,m}$ the input–output coefficient matrix of material cp in process $m \in M_{pet}$, and $D_{Pet\ cp}^{L}$ and $D_{Pet\ cp}^{U}$ represent the lower and upper level of demand for product $cp \in CP$, respectively. Then, the material balance that governs the operation of the petrochemical network can be expressed as shown in constraints (4.1) and (4.2):

$$F_{cp}^{Pet} + \sum_{m \in M_{Pet}} \delta_{cp,m}\, x_m^{Pet} \geq D_{Pet\ cp}^{L} \quad \forall\, cp \in CP \tag{4.1}$$

$$F_{cp}^{Pet} + \sum_{m \in M_{Pet}} \delta_{cp,m}\, x_m^{Pet} \leq D_{Pet\ cp}^{U} \quad \forall\, cp \in CP \tag{4.2}$$

For a given subset of chemicals, where $cp \in CP$, these constraints control the production of different processes based on the upper and lower demands of the petrochemical market for the final products. In constraint (4.3), defining the binary variables $y_{proc\ m}^{Pet}$ for each process $m \in M_{Pet}$ is required for the process selection requirement as $y_{proc\ m}^{Pet}$ will equal 1 only if process m is selected or zero otherwise. Furthermore, if only process m is selected, its production level must be at least equal to the process minimum economic capacity B_m^L for each $m \in M_{Pet}$, where K^U is a valid upper bound.. This can be written for each process m as follows:

$$B_m^L\, y_{proc\ m}^{Pet} \leq x_m^{Pet} \leq K^U\, y_{proc\ m}^{Pet} \quad \forall\, m \in M_{Pet} \tag{4.3}$$

In the case where it is preferred to choose only one process technology to produce a single chemical, constraints (4.4) and (4.5) can be included for each intermediate and product chemical type, respectively:

$$\sum_{cp \in CIP} y^{Pet}_{proc\ m} \leq 1 \quad \forall\, m \in M_{Pet} \text{ that produces } cp \in CIP \text{ (intermediate)} \quad (4.4)$$

$$\sum_{cp \in CFP} y^{Pet}_{proc\ m} \leq 1 \quad \forall\, m \in M_{Pet} \text{ that produces } cp \in CFP \text{ (final)} \quad (4.5)$$

Finally, we can specify limitations on the supply of feedstock S^{Pet}_{cp} for each chemical type cp though constraint (4.6):

$$F^{Pet}_{cp} \leq S^{Pet}_{cp} \quad \forall\, cp \in CP \quad (4.6)$$

The economic objective in the model can either be represented as operating cost minimization or added-value maximization. In the case of added-value maximization, product prices are subtracted from the cost of feedstocks for each process. If Pr^{Pet}_{cp} is the price of chemical cp, the added-value objective function can be represented as:

$$\text{Max} \sum_{cp \in CP} \sum_{m \in M_{Pet}} Pr^{Pet}_{cp}\, \delta_{cp,m}\, x^{Pet}_{m} \quad (4.7)$$

4.4
Illustrative Case Study

The case study presented in this book is based on Al-Sharrah, Hankinson and Elkamel (2006). The petrochemical network included 81 processes connecting the production and consumption of 65 chemicals. Simplified networks of processes and chemicals included in the petrochemical network are given in Figure 4.1 and Table 4.1; respectively. The chemicals are classified according to their function as follows:

1) Primary raw material (PR)
2) Secondary raw material (SR)
3) Intermediate (I)
4) Primary final product (PF)
5) Secondary final product (SF).

PR chemicals are derived from petroleum and natural gas and other basic feedstocks, whereas the SR chemicals are those needed as additives or in small quantities.

The chemicals classified as I are those produced and consumed in the petrochemical network. Finally, the PF and SF chemicals are the selected final products by selected processes and the associated byproducts in the network, respectively.

The modeling system GAMS (Brooke et al., 1996) is used for setting up the optimization models. The computational tests were carried out on a Pentium M

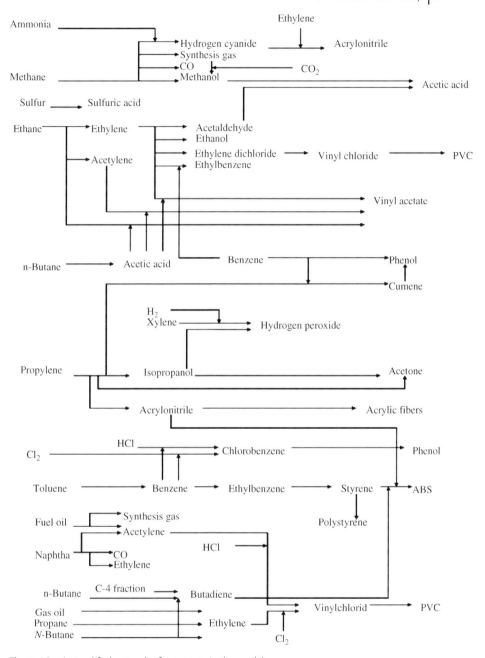

Figure 4.1 A simplified network of processes in the model.

Table 4.1 A list of chemicals included in the model.

Chemical	Function
Acetaldehyde	SF + I
Acetic acid	I + PF
Acetone	PF
Acetylene	I
Acrylic fibers	PF
Acrylonitrile	I
Acrylonitrile–butadiene–styrene	PF
Ammonia	PR
Benzene	SF + I
Butadiene	I
Butenes (mixed *n*-, iso-,-dienes, etc.)	SF + PR
C-4 fraction (mixed butanes, -enes, etc.)	SF + PR
Carbon dioxide	SR
Carbon monoxide	I
Chlorine	PR
Chlorobenzene	I
Coke	PR
Cumene	I + PF
Ethane	PR
Ethanol	I
Ethyl benzene	I
Ethylene	SF + I
Ethylene dichloride	I
Formic acid	SF
Fuel gas	SF
Fuel oil	SF + PR
Gas oil	PR
Gasoline	SF
Hydrochloric acid	SR
Hydrogen	SR + SF
Hydrogen cyanide	I
Hydrogen peroxide	I
Isopropyl alcohol	I
Methane	PR + SF
Methanol	I
Methyl acrylate	SR
Methyl methacrylate	SR
Naphtha	PR
n-Butane	PR
n-Butylenes (1- and 2-)	PR
Pentane	SR
Phenol	PR
Polybutadiene rubber	SR
Polystyrene (crystal grade)	I + PF
Polystyrene (expandable beads)	PF
Polystyrene (impact grade)	PF

Table 4.1 (Continued)

Chemical	Function
Poly(vinyl acetate)	I
Poly(vinyl alcohol)	SR
Poly(vinyl chloride)	PF
Propane	SF + PR
Propylene (chemical grade)	SF + I
Propylene (refinery grade)	PR
Propylene oxide	SF
Sodium carbonate	SR
Sodium hydroxide	SR
Styrene	I
Sulfuric acid	I
Sulfur	PR
Synthesis gas 3:1	I
Synthesis gas 2:1	SF
Toluene	PR + SF
Vinyl acetate	I + PF
Vinyl chloride	I
Xylene (mixed)	SR + SF

The potential function of each chemical is also indicated: PR = primary raw material, SR = secondary raw material, I = intermediate, PF = primary final product, SF = secondary final product.

processor 2.13 GHz. The MILP model was solved using CPLEX (CPLEX Optimization Inc., 1993).

The model in this form is moderate in size and the solution indicated the selection of 22 processes out of the 81 processes proposed. The selected processes and their respective capacities are shown in Table 4.2. This case study represents an ideal situation where all parameters are known with certainty.

The final petrochemical network suggests the use of lighter petroleum refining feedstocks. The petrochemical network mainly used ethane, propane, C-4 fractions (mixed butanes, -enes, etc.), pentane, and refinery grade propylene. In the case of lower lighter petroleum product availability, the network will suggest the use of steam cracking of naphtha or gas oil. This will be required in order to obtain the main petrochemical building blocks for the downstream processes that include ethylene and chemical grade propylene. The annual production benefit of the petrochemical network was found to be $ 2 202 268.

4.5
Conclusion

In this chapter we presented an MILP deterministic planning model for the optimization of a petrochemical network. The optimization model presents a tool

Table 4.2 Deterministic model solution.

Process selected	Production Capacity (10^3 t/year)
acetaldehyde by the one-step oxidation from ethylene	1015.5
acetic acid by air oxidation of acetaldehyde	404.6
acetone by oxidation of propylene	169.8
acetylene by submerged flame process	179.8
acrylic fibers by batch suspension polymerization	246
acrylonitrile by cyanation/oxidation of ethylene	294.9
ABS by suspension/emulsion polymerization	386.9
benzene by hydrodealkylation of toluene	432.3
butadiene by extractive distillation	96.7
chlorobenzene by oxychlorination of benzene	73.0
cumene by the reaction of benzene and propylene	72.0
ethylbenzene by the alkylation of benzene	458.8
ethylene by steam cracking of ethane–propane (50–50 wt%)	1068.3
hydrogen cyanide by the ammoxidation of methane	177.0
phenol by dehydrochlorination of chlorobenzene	61.4
polystyrene (crystal grade) by bulk polymerization	66.8
polystyrene (expandable beads) by suspension polymerization	51.5
polystyrene (impact grade) by suspension polymerization	77.1
poly(vinyl chloride) by bulk polymerization	408.0
styrene from dehydrogenation of ethylbenzene	400.0
vinyl acetate from reaction of ethane and acetic acid	113.9
vinyl chloride by the hydrochlorination of acetylene	418.2

that simplifies the process of decision-making for such large and complex petrochemical systems.

However, considering this type of high level strategic planning model, especially with the current volatile market environment and the continuous change in customer requirements, the impact of uncertainties is inevitable. In fact, ignoring uncertainty of key parameters in decision problems can yield non-optimal and infeasible decisions (Birge, 1995). For this reason, the scope of the next section of this book is to extend the deterministic petrochemical planning model to account for uncertainties in model parameters and include the risk notion in decision-making using proper robust optimization techniques.

References

Al-Amir, A.M.J., Al-Fares, H. and Rahman, F. (1998) Towards modeling the development of the petrochemical industry in the kingdom of Saudi Arabia. *Arabian Journal of Science & Engineering*, **23**, 113.

Al-Fares, H.K. and Al-Amir, A.M. (2002) An optimization model for guiding the petrochemical industry development in Saudi Arabia. *Engineering Optimization*, **34**, 671.

References

Al-Sharrah, G.K., Hankinson, G. and Elkamel, A. (2006) Decision-making for petrochemical planning using multiobjective and strategic tools. *Chemical Engineering Research and Design*, **84** (A11), 1019.

Al-Sharrah, G.K., Alatiqi, I. and Elkamel, A. (2002) Planning an integrated petrochemical business portfolio for long-range financial stability. *Industrial & Engineering Chemistry Research*, **41**, 2798.

Al-Sharrah, G.K., Alatiqi, I. and Elkamel, A. (2003) Modeling and identification of economic disturbances in the planning of the petrochemical industry. *Industrial & Engineering Chemistry Research*, **42**, 4678.

Al-Sharrah, G.K., Alatiqi, I., Elkamel, A. and Alper, E. (2001) Planning an integrated petrochemical industry with an environmental objective. *Industrial & Engineering Chemistry Research*, **40**, 2103.

Birge, J.R. (1995) Models and model value in stochastic programming. *Annals of Operations Research*, **59**, 1.

Brooke, A., Kendrick, D., Meeraus, A. and Raman, R. (1996) *GAMS–A User's Guide*, GAMS Development Corporation, Washington DC.

CPLEX Optimization, Inc. (1993) *Using the CPLEX Callable Library and CPLEX Mixed Integer Library*, CPLEX Optimization, Inc., Incline Village, NV.

Energy information administration (EIA) (2006) *International Energy Outlook*. http://www.eia.doe.gov/oiaf/ieo/index.html, accessed on October 20, 2006.

Fathi-Afshar, S. and Yang, J. (1985) Designing the optimal structure of the petrochemical industry for minimum cost and least gross toxicity of chemical production. *Chemical Engineering Science*, **40**, 781.

Jimenez, A. and Rudd, D.F. (1987) Use of a recursive mixed-integer programming model to detect an optimal integration sequence for the mexican petrochemical industry. *Computers & Chemical Engineering*, **11**, 291.

Jimenez, A., Rudd, D.F. and Meyer, R.R. (1982) Study of the development of mexican petrochemical industry using mixed-integer programming. *Computers & Chemical Engineering*, **6**, 219.

Sokic, M. and Stevancevic, D. (1983) The optimal structure of the petrochemical industry. *Chemical Engineering Science*, **38**, 265.

Sophos, A., Rotstein, E. and Stephanopoulos, G. (1980) Multiobjective analysis in modeling the petrochemical industry. *Chemical Engineering Science*, **38**, 2415.

Stadtherr, M.A. and Rudd, D.F. (1976) Systems study of the Petrochemical Industry. *Chemical Engineering Science*, **31**, 1019.

Stadtherr, M.A. and Rudd, D.F. (1978) Resource use by the petrochemical industry. *Chemical Engineering Science*, **33**, 923.

5
Multisite Refinery and Petrochemical Network Integration

Achieving optimal integration between refineries and the petrochemical industry was, until recently, a far-fetched target. Now many refiners are engaging in feasibility studies to include and expand their share in the petrochemical market.

In this chapter, we present a model for the design of optimal integration and coordination of a refinery and petrochemical network to satisfy given demand for chemical products. The main feature of the approach is the development of a methodology for the simultaneous analysis of process network integration within a multisite refinery and petrochemical system, thus achieving a global optimal production strategy by allowing appropriate trade-offs between the refinery and the downstream petrochemical markets. The performance of the proposed model is tested on industrial-scale examples of multiple refineries and a polyvinyl chloride (PVC) complex to illustrate the economic potential and trade-offs involved in the optimization of the network.

5.1
Introduction

In view of the current situation of high oil prices and the increasing consciousness and implementation of strict environmental regulations, petroleum refiners and petrochemical companies have started to seek opportunities for mergers and integration. This is evident in the current projects around the world for building integrated refineries and the development of complex petrochemical industries that are aligned through advanced integration platforms. Figure 5.1 illustrates a typical refining and petrochemical industry supply chain. The realization of coordination and objective alignment benefits across the enterprise has been the main driver of such efforts (Sahinidis *et al.*, 1989; Shah, 1998).

Despite the fact that petroleum refining and petrochemical companies have recently engaged in more integration projects, relatively little research has been reported in the open literature, mostly due to confidentiality reasons. Such concerns render the development of a systematic framework of network integration and coordination

Planning and Integration of Refinery and Petrochemical Operations. Khalid Y. Al-Qahtani and Ali Elkamel
Copyright © 2010 WILEY-VCH Verlag GmbH & Co. KGaA, Weinheim
ISBN: 978-3-527-32694-5

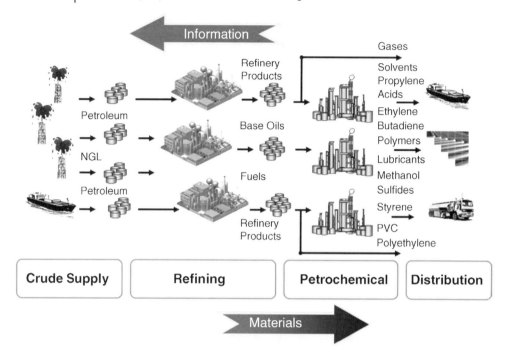

Figure 5.1 Refinery and petrochemical industry supply chain.

difficult. The studies published in the open literature were mainly developed by consulting and design firms, as well as the operating companies, and generally lack a structured methodology for evaluating the projects feasibility. Just to mention a few of the studies published, Swaty (2002) studied the possibility of integrating a refinery and an ethylene plant through the exchange of process intermediate streams. The analysis was based on a linear programming model for each plant and profit marginal analysis of possible intermediate plant exchange. The study was implemented on a real life application in western Japan. Gonzalo et al. (2004) highlighted the benefits of refining and petrochemical integration. They discussed a project dealing with the installation of a hydrocracker in Repsol's refinery in Spain and how it improved the overall synergy between the refinery and a stream cracker plant.

In the academic arena, Sadhukhan, Zhang, and Zhu (2004) developed an analytical flowsheet optimization method for applications in the petroleum refining and petrochemical industry. The proposed methodology consisted of three main steps: market integration, facility network optimization through economic margin analysis and load shifting, and elimination of less profitable processes. They demonstrated their method on two case studies: a single site refinery and a petrochemical complex. Li et al. (2006) proposed a linear programming (LP) model for the integration of a refinery and an ethylene cracker. They evaluated different schemes iteratively for different crude types to optimize the refinery and ethylene cracker operations. The best scheme was selected based on the highest profit from the cases studied. More recently, Kuo and Chang (2008) developed a short term multi-period planning model for a benzene–toluene–xylene (BTX) complex. They modeled the system as a mixed-integer linear programming (MILP) model with binary variables mainly for mode switching and inventory control (backlog and surplus indicators). They divided the processing units into two sets, (i) reaction processes: reforming, isomerization units and tatory units, and (ii) separation processes: aromatics extraction units, xylene fractionation units and the parex units. Optimization of the other refinery units, blending levels and olefin cracking processes were not considered. The decisions in their study mainly included the optimal throughput and operation mode of each production unit, inventory levels and feedstock supplies. For more details on strategic multisite planning studies, multisite refinery optimization, and petrochemical industry planning the interested reader is referred to the previous chapters and the references given herein.

The integration of petroleum refining and petrochemical plants is gaining a great deal of interest with the realization of the benefits of coordination and vertical integration. Previous research in the field assumed either no limitations on refinery feedstock availability for the petrochemical planning problem or fixed the refinery production levels assuming an optimal operation. However, in this chapter we present a mathematical model for the determination of the optimal integration and coordination strategy for a refinery network and synthesize the optimal petrochemical network required to satisfy a given demand from any set of available technologies. Therefore, achieving a global optimal production strategy by allowing appropriate trade-offs between the refinery and the downstream petrochemical markets. The refinery and petrochemical systems were modeled as MILP problems that will also lead to overall refinery and petrochemical process production levels and detailed blending levels at each refinery site. The objective function is a minimization of the annualized cost over a given time horizon among the refineries by improving the coordination and utilization of excess capacities in each facility and maximization of the added value in the petrochemical system. Expansion requirements to improve production flexibility and reliability in the refineries are also considered.

The remainder of Chapter 5 is organized as follows. In Section 5.2, we explain the problem statement of the petroleum refinery and petrochemical integration. Then, we discuss the proposed model formulation in Section 5.3. In Section 5.4, we illustrate the performance of the model through an industrial-scale case study. The chapter ends with some concluding remarks in Section 5.5.

5.2
Problem Statement

The optimization of refining and petrochemical processes involves a wide range of aspects, varying from economical analysis and strategic expansions to crude oil selection, process levels targets, operating modes, and so on. The focus of this chapter is to develop a mathematical programming tool for the simultaneous design of an integrated network of refineries and petrochemical processes. On the refinery side, the model provides the optimal network integration between the refineries, process expansion requirements, operating policy based on different feedstock combination alternatives, process levels and operating modes. On the petrochemical side, the model establishes the design of an optimal petrochemical process network from a range of process technologies to satisfy a given demand. The simultaneous network design and optimization of the refining and petrochemical industry provides appropriate means for improving the coordination across the industrial system and can develop an overall optimal production strategy across the petroleum chain.

The general problem under study can be defined as follows: A set of refinery products $cfr \in CFR$ produced at multiple refinery sites $i \in I$ and a set of petrochemical products $cp \in CFP$ is given. Each refinery consists of different production units $m \in M_{Ref}$ that can operate at different modes $p \in P$ while a set of wide range petrochemical and chemical process technologies $m \in M_{Pef}$ is available for selection. Furthermore, different crude oil slates $cr \in CR$ are available and given. The petrochemical network selects its feedstock from three main sources; namely, refinery intermediate streams $Fi^{Pet}_{cr,cir,i}$ of an intermediate product $cir \in RPI$, refinery final products $Ff^{Pet}_{cr,cfr,i}$ of a final product $cfr \in RPF$, and non-refinery streams Fn^{Pet}_{cp} of a chemical $cp \in NRF$. The process network across the refineries and petrochemical system is connected in a finite number of ways. An integration superstructure between the refinery processes is defined in order to allow exchanging intermediate streams. Market product prices, operating cost at each refinery and petrochemical system, as well as product demands are assumed to be known.

The problem is to determine the optimal integration and coordination strategy for the overall refinery network and to design the optimal petrochemical network required to satisfy a given demand from the available process technologies. The proposed approach will also provide overall refinery and petrochemical process production levels and detailed blending levels at each refinery site. The objective function is a minimization of the annualized cost over a given time horizon among the refineries by improving the coordination and utilization of excess capacities in each facility and maximization of the added value in the petrochemical system. Expansion requirements to improve production flexibility and reliability in the refineries are also considered.

For all refinery and petrochemical processes within the network we assume that all material balances are expressed in terms of linear yield vectors. Even though this might sound restrictive, as most if not all refinery and petrochemical processes are inherently nonlinear, this practice is commonly applied with such large scale systems. Moreover, the decisions in this study are of a strategic level in which such linear formulation is adequate to address the required level of detail involved at this stage. It is also assumed that processes have fixed capacities and the operating cost of each process and production mode is proportional to the process inlet flow. In the case of refinery product blending, quality blending indices are used to maintain model linearity. It is also assumed that all products that are in excess of the local demand can be exported to a global market. Piping and pumping installation costs to transport refinery intermediate streams between the different refinery sites as well as the operating costs of the new system are lumped into one fixed-charge cost. All costs are discounted over a 20 years time horizon and with an interest rate of 7%. No inventories will be considered since the model is addressing strategic decisions which usually cover a long period of time. We also assume perfect mixing in the refineries and that the properties of each crude oil slate are decided by specific key components.

5.3 Model Formulation

The proposed formulation addresses the problem of simultaneous design of an integrated network of refineries and petrochemical processes. The proposed model is based on the formulations proposed in this dissertation. All material balances are carried out on a mass basis with the exception of refinery quality constraints of properties that only blend by volume where volumetric flow rates are used instead. The model is formulated as follows:

$$\text{Min} \sum_{cr \in CR} \sum_{i \in I} CrCost_{cr} \, S^{Ref}_{cr,i}$$

$$+ \sum_{p \in P} OpCost_p \sum_{cr \in CR} \sum_{i \in I} z_{cr,p,i}$$

$$+ \sum_{cir \in CIR} \sum_{i \in I} \sum_{i' \in I} InCost_{i,i'} \, Ypipe^{Ref}_{cir,i,i'} \qquad \text{where} \quad i \neq i' \qquad (5.1)$$

$$+ \sum_{i \in I} \sum_{m \in M_{Ref}} \sum_{s \in S} InCost_{m,s} \, Yexp^{Ref}_{m,i,s}$$

$$- \sum_{cfr \in PEX} \sum_{i \in I} Pr^{Ref}_{cfr} \, e^{Ref}_{cfr,i}$$

$$- \sum_{cp \in CP} \sum_{m \in M_{Pet}} Pr^{Pet}_{cp} \, \delta_{cp,m} \, x^{Pet}_m$$

Subject to

$$z_{cr,p,i} = S^{Ref}_{cr,i} \quad \begin{array}{l} \forall \, cr \in CR, i \in I \text{ where} \\ p \in P' = \{\text{Set of CDU processes } \forall \text{ plant } i\} \end{array} \qquad (5.2)$$

$$\sum_{p \in P} \alpha_{cr,cir,i,p} \, z_{cr,p,i} + \sum_{i' \in I} \sum_{p \in P} \xi_{cr,cir,i',p,i} \, xi^{Ref}_{cr,cir,i',p,i} - Fi^{Pet}_{cr,cir \in RPI,i}$$

$$- \sum_{i' \in I} \sum_{p \in P} \xi_{cr,cir,i,p,i'} \, xi^{Ref}_{cr,cir,i,p,i'} - \sum_{cfr \in CFR} w_{cr,cir,cfr,i} - \sum_{rf \in FUEL} w_{cr,cir,rf,i} = 0$$

$$\forall \quad \begin{array}{l} cr \in CR, \\ cir \in CIR, \\ i' \text{ and } i \in I \end{array} \quad \text{where} \quad i \neq i' \qquad (5.3)$$

$$\sum_{cr \in CR}\sum_{cir \in CB} w_{cr,cir,cfr,i} - \sum_{cr \in CR}\sum_{rf \in FUEL} w_{cr,cfr,rf,i} - \sum_{cr \in CR} Ff^{Pet}_{cr,cfr \in RPF,i} = x^{Ref}_{cfr,i} \quad \forall \quad \begin{array}{l} cfr \in CFR, \\ i \in I \end{array} \tag{5.4}$$

$$\sum_{cr \in CR}\sum_{cir \in CB} \frac{w_{cr,cir,cfr,i}}{SG_{cr,cir}} = xv^{Ref}_{cfr,i} \quad \forall \quad \begin{array}{l} cfr \in CFR, \\ i \in I \end{array} \tag{5.5}$$

$$\sum_{cir \in FUEL} cv_{rf,cir,i}\, w_{cr,cir,rf,i} + \sum_{cfr \in FUEL} w_{cr,cfr,rf,i} - \sum_{p \in P} \beta_{cr,rf,i,p}\, z_{cr,p,i} = 0 \quad \forall \quad \begin{array}{l} cr \in CR, \\ rf \in FUEL, \\ i \in I \end{array} \tag{5.6}$$

$$\sum_{cr \in CR}\sum_{cir \in CB} \left(\begin{array}{c} att_{cr,cir,q \in Qv}\, \dfrac{w_{cr,cir,cfr,i}}{SG_{cr,cir}} + att_{cr,cir,q \in Qw} \\[4pt] \left[w_{cr,cir,cfr,i} - \sum_{rf \in FUEL} w_{cr,cfr,rf,i} - \sum_{cr \in CR} Ff^{Pet}_{cr,cfr \in RPF,i} \right] \end{array} \right) \geq q^{L}_{cfr,q \in Qv}\, xv^{Ref}_{cfr,i} + q^{L}_{cfr,q \in Qw}\, x^{Ref}_{cfr,i} \quad \forall \quad \begin{array}{l} cfr \in CFR, \\ q = \{Qw, Qv\}, \\ i \in I \end{array} \tag{5.7}$$

$$\sum_{cr \in CR}\sum_{cir \in CB} \left(\begin{array}{c} att_{cr,cir,q \in Qv}\, \dfrac{w_{cr,cir,cfr,i}}{SG_{cr,cir}} + att_{cr,cir,q \in Qw} \\[4pt] \left[w_{cr,cir,cfr,i} - \sum_{rf \in FUEL} w_{cr,cfr,rf,i} - \sum_{cr \in CR} Ff^{Pet}_{cr,cfr \in RPF,i} \right] \end{array} \right) \leq q^{U}_{cfr,q \in Qv}\, xv^{Ref}_{cfr,i} + q^{U}_{cfr,q \in Qw}\, x^{Ref}_{cfr,i} \quad \forall \quad \begin{array}{l} cfr \in CFR, \\ q = \{Qw, Qv\}, \\ i \in I \end{array} \tag{5.8}$$

$$\text{Min } C_{m,i} < \sum_{p \in P} \gamma_{m,p} \sum_{cr \in CR} z_{cr,p,i} < \text{Max } C_{m,i} + \sum_{s \in S} AddC_{m,i,s}\, y^{Ref}_{exp_{m,i,s}} \quad \forall\, m \in M_{Ref}, i \in I \tag{5.9}$$

$$\sum_{cr \in CR}\sum_{p \in P} \xi_{cr,cir,i,p,i'}\, xi^{Ref}_{cr,cir,i,p,i'} \leq F^{U}_{cir,i,i'}\, y^{Ref}_{pipe_{cir,i,i'}} \quad \forall\, cir \in CIR, i' \text{ and } i \in I \text{ where } i \neq i' \tag{5.10}$$

$$\sum_{i \in I} \left(x^{Ref}_{cfr,i} - e^{Ref}_{cfr',i} \right) \geq D^{Ref}_{cfr} \quad \forall\, cfr \text{ and } cfr' \text{ where } cfr \in CFR, cfr' \in PEX \tag{5.11}$$

$$IM^{L}_{cr} \leq \sum_{i \in I} S^{Ref}_{cr,i} \leq IM^{U}_{cr} \quad \forall\, cr \in CR \tag{5.12}$$

$$Fn^{Pet}_{cp \in NRF} + \sum_{i \in I} \sum_{cr \in CR} Fi^{Pet}_{cr,cp \in RPI,i} + \sum_{i \in I} \sum_{cr \in CR} Ff^{Pet}_{cr,cp \in RPF,i}$$
$$+ \sum_{m \in M_{Pet}} \delta_{cp,m}\, x^{Pet}_m \geq D^L_{Pet\ cp \in CFP} \qquad \forall\, cp \in CP \quad (5.13)$$

$$Fn^{Pet}_{cp \in NRF} + \sum_{i \in I} \sum_{cr \in CR} Fi^{Pet}_{cr,cp \in RPI,i} + \sum_{i \in I} \sum_{cr \in CR} Ff^{Pet}_{cr,cp \in RPF,i}$$
$$+ \sum_{m \in M_{Pet}} \delta_{cp,m}\, x^{Pet}_m \leq D^U_{Pet\ cp \in CFP} \qquad \forall\, cp \in CP \quad (5.14)$$

$$B^L_m\, y_{proc^{Pet}_m} \leq x^{Pet}_m \leq K^U\, y_{proc^{Pet}_m} \quad \forall\, m \in M_{Pet} \quad (5.15)$$

$$\sum_{cp \in CIP} y_{proc^{Pet}_m} \leq 1 \quad \forall\, m \in M_{Pet}\ \text{that produces}\ cp \in CIP \quad (5.16)$$

$$\sum_{cp \in CFP} y_{proc^{Pet}_m} \leq 1 \quad \forall\, m \in M_{Pet}\ \text{that produces}\ cp \in CFP \quad (5.17)$$

$$Fn^{Pet}_{cp} \leq S^{Pet}_{cp} \quad \forall\, cp \in NRF \quad (5.18)$$

The above objective function (5.1) represents a minimization of the annualized cost which consists of crude oil cost, refineries operating cost, refineries intermediate exchange piping cost, refinery production system expansion cost, less the refinery export revenue and added value by the petrochemical processes. The operating cost of each refinery process is assumed to be proportional to the process inlet flow and is expressed on a yearly basis. Inequality (5.2) corresponds to each refinery raw materials balance where throughput to each distillation unit $p \in P'$ at plant $i \in I$ from each crude type $cr \in CR$ is equal to the available supply $S^{Ref}_{cr,i}$. Constraint (5.3) represents the intermediate material balances within and across the refineries where the coefficient $\alpha_{cr,cir,i,p}$ can assume either a positive sign if it is an input to a unit or a negative sign if it is an output from a unit. The multi-refinery integration matrix $\xi_{cr,cir,i,p,i'}$ accounts for all possible alternatives of connecting intermediate streams $cir \in CIR$ of crude $cr \in CR$ from refinery $i \in I$ to process $p \in P$ in plant $i' \in I'$. Variable $xi^{Ref}_{cr,cir,i,p,i'}$ represents the trans-shipment flow rate of crude $cr \in CR$, of intermediate $cir \in CIR$ from plant $i \in I$ to process $p \in P$ at plant $i' \in I$. Constraint (5.3) also considers the petrochemical network feedstock from the refinery intermediate streams $Fi^{Pet}_{cr,cir,i}$ of each intermediate product $cir \in RPI$. The material balance of final products in each refinery is expressed as the difference between flow rates from intermediate streams $w_{cr,cir,cfr,i}$ for each $cir \in CIR$ that contribute to the final product pool and intermediate streams that contribute to the fuel system $w_{cr,cfr,rf,i}$ for each $rf \in FUEL$ less the refinery final products $Ff^{Pet}_{cr,cfr,i}$ for each $cfr \in RPF$ that are fed to the petrochemical network, as shown in constraint (5.4). In constraint (5.5) we convert the mass flow rate to volumetric flow rate by dividing it by the specific gravity $SG_{cr,cir}$ of each crude type $cr \in CR$ and intermediate stream $cir \in CB$. This is needed in order to express the quality attributes that blend by volume in blending pools. Constraint (5.6) is the fuel system material balance where the term $cv_{rf,cir,i}$ represents the caloric value equivalent for each intermediate $cir \in CB$ used in the fuel system at plant $i \in I$. The fuel production system can consist of either a single or combination of intermediates $w_{cr,cir,rf,i}$ and products $w_{cr,cfr,rf,i}$. The matrix $\beta_{cr,rf,i,p}$ corresponds to the consumption of each processing unit $p \in P$ at plant $i \in I$ as a percentage of unit throughput. Constraints (5.7) and (5.8), respectively, represent lower and upper bounds on refinery quality constraints for all refinery products that either blend by mass $q \in Q_w$ or by volume $q \in Q_v$. Constraint (5.9) represents the maximum and minimum allowable flow rate to each processing unit. The coefficient $\gamma_{m,p}$ is a zero-one matrix for the assignment of production unit $m \in M_{Ref}$ to process operating mode $p \in P$. The term $AddC_{m,i,s}$ accounts for the additional refinery expansion capacity of each production unit $m \in M_{Ref}$ at refinery $i \in I$ for a specific expansion size $s \in S$. In this formulation, we only allow the addition of predetermined capacities whose pricing can be acquired a priori through design companies' quotations. The integer variable $\gamma\mathrm{exp}_{m,i,s}^{Ref}$ represents the decision of expanding a production unit and it can take a value of one if the unit expansion is required or zero otherwise. Constraint (5.10) sets an upper bound on the flow rates of intermediate streams between the different refineries. The integer variable $\gamma\mathrm{pipe}_{cir,i,i'}^{Ref}$ represents the decision of exchanging intermediate products between the refineries and takes on the value of one if the commodity is transferred from plant $i \in I$ to plant $i' \in I$ or zero otherwise, where $i \neq i'$. When an intermediate stream is selected to be exchanged between two refineries, its flow rate must be below the transferring pipeline capacity $F_{cir,i,i'}^U$. Constraint (5.11) stipulates that the final products from each refinery $x_{cfr,i}^{Ref}$ less the amount exported $e_{cfr',i}^{Ref}$ for each exportable product $cfr' \in PEX$ from each plant $i \in I$ must satisfy the domestic demand $D_{Ref_{cfr}}$. Resources are limited by constraint (5.12) which imposes upper and lower bounds on the available feedstock $cr \in CR$ to the refineries.

Constraints (5.13) and (5.14) represent the material balance that governs the operation of the petrochemical system. The variable x_m^{Pet} represents the annual level of production of process $m \subset M_{Pet}$ where $\delta_{cp,m}$ is the input–output coefficient matrix of material cp in process $m \in M_{Pet}$. The petrochemical network receives its feed from potentially three main sources. These are, (i) refinery intermediate streams $Fi_{cr,cir,i}^{Pet}$ of an intermediate product $cir \in RPI$, (ii) refinery final products $Ff_{cr,cfr,i}^{Pet}$ of a final product $cfr \in RPF$, and (iii) non-refinery streams Fn_{cp}^{Pet} of a chemical $cp \in NRF$. For a given subset of chemicals $cp \in CP$, the proposed model selects the feed types, quantity and network configuration based on the final chemical and petrochemical lower and upper product demand $D\mathrm{Pet}_{cp}^L$ and $D\mathrm{Pet}_{cp}^U$ for each $cp \in CFP$, respectively. In constraint (5.15), defining a binary variable $\gamma\mathrm{proc}_m^{Pet}$ for each process $m \in M_{pet}$ is required for the process selection requirement as $\gamma\mathrm{proc}_m^{Pet}$ will equal 1 only if process m is selected or zero otherwise. Furthermore, if only process m is selected, its production level must be at least equal to the process minimum economic capacity B_m^L for each $m \in M_{pet}$, where K^U is a valid upper bound. In the case where it is preferred to choose only one process technology to produce a chemical, constraints (5.16) and (5.17) can be included for each intermediate product $cp \in CIP$ and final product $cp \in CFP$, respectively. Finally, we can specify limitations on the supply of feedstock Fn_{cp}^{Pet} for each chemical type $cp \in NRF$ through constraint (5.18). Bear in mind that the limitations on the refinery intermediate product $Fi_{cr,cir,i}^{Pet}$ and final product $Ff_{cr,cfr,i}^{Pet}$ that are fed to the petrochemical network are dictated by the model based on both refinery and petrochemical demand and price structure.

5.4
Illustrative Case Study

In this section we demonstrate the performance of the proposed model on an industrial-scale case study. Instead of considering the full scale petrochemical network which may have limited application, we consider a special case of the integration problem. Although the proposed formulation covers the full scale refinery network and petrochemical systems, the case study will consider the integration of a petrochemical complex for the production of polyvinyl chloride (PVC) with a multi-refinery network. PVC is one of the major ethylene derivatives that has many important applications and uses, including pipe fittings, automobile bumpers, toys, bottles and many others (Rudd et al., 1981).

Direct integration of refining and ethylene cracking is considered as the essential building block in achieving the total petrochemical integration (Joly, Moro, and Pinto, 2002; Li et al., 2006). This problem has received more attention lately due to soaring motor gasoline prices and the directly related prices of ethylene feedstocks (Lippe, 2007). This kind of volatility in prices has prompted a shift in ethylene feedstock selection and economics to either lighter or heavier refinery product slates (Lippe, 2008). Shifting from one feedstock to another will mainly depend on the market price structure and demand for refinery products. Some researchers believe that the tendency of ethylene feedstock shift would mainly be towards heavier refinery streams including heavy and vacuum gas oils due to the diminishing reserves of sweet crudes and decreasing demand for heavy fraction fuels (Singh et al., 2005; Van Geem et al., 2008). This change in ethylene feedstock selection and the direct effect on the refinery products requires adequate decision making and analysis that takes into account both refining and petrochemical markets.

In the case study, we consider the planning for three complex refineries by which we demonstrate the performance of our model in devising an overall production plan as well as an integration strategy among the refineries. The state equipment network (SEN) representation for the overall topology of the refineries network is given in Figure 5.2.

The atmospheric crude unit separates crude oil into several fractions including LPG, naphtha, kerosene, gas oil and residues. The heavy residues are then sent to the vacuum unit where they are further separated into vacuum gas oil and vacuum

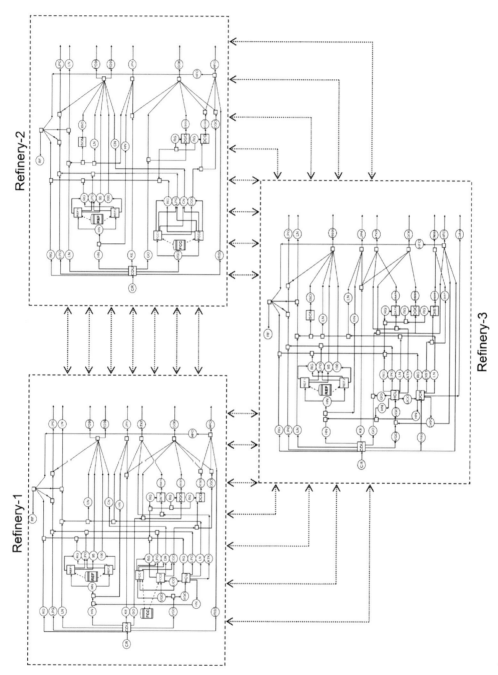

Figure 5.2 SEN representation of the refinery integration network.

residues. Depending on the production targets, different processing and treatment processes are applied to the crude fractions. In this example, the naphtha is further separated into heavy and light naphtha. Heavy naphtha is sent to the catalytic reformer unit to produce high octane reformates for gasoline blending and light naphtha is sent to the light naphtha pool and to an isomerization unit to produce isomerate for gasoline blending too, as in refineries 2 and 3. The middle distillates are combined with other similar intermediate streams and sent for hydrotreating and then for blending to produce jet fuels and gas oils. Atmospheric and vacuum gas oils are further treated by catalytic cracking, as in refinery 2, or by hydrocracking, as in refinery 3, or by both, as in refinery 1, to increase the gasoline and distillate yields. In refinery 3, vacuum residues are further treated using coking and thermal processes to increase light products yields. The final products of the three refineries network consists of liquefied petroleum gas (LPG), light naphtha (LT), two grades of gasoline (PG98 and PG95), no. 4 jet fuel (JP4), military jet fuel (ATKP), no. 6 gas oil (GO6), diesel fuel (Diesel), heating fuel oil (HFO), and petroleum coke (coke). The major capacity constraints for the refinery network are given in Table 5.1. Furthermore, the

Table 5.1 Major refinery network capacity constraints.

Production Capacity	Higher limit (10^3 t/year)		
	R1	R2	R3
Distillation	45 000.	12 000.0	9900.0
Reforming	700.0	2000.0	1800.0
Isomerization	200.0	—	450.0
Fluid catalytic cracker	800.0	1400.0	—
Hydrocracker	—	1800.0	2400.0
Delayed coker	—	—	1800
Des gas oil	1300.0	3000.0	2400.0
Des cycle gas oil	200.0	750.0	—
Des ATK	—	1200.0	1680.0
Des distillates	—	—	450.0
Crude availability			
Arabian light		31 200.0	
Local Demand			
LPG		432.0	
LN		—	
PG98		540.0	
PG95		4440.0	
JP4		2340.0	
GO6		4920.0	
ATK		1800.0	
HFO		200.0	
Diesel		400.0	
Coke		300.0	

three refineries are assumed to be in one industrial area, which is a common situation in many locations around the world, and are coordinated through a main headquarters sharing the feedstock supply. The cost parameters for pipelines installation were calculated as cost per distance between the refineries, and then multiplied by the required pipe length in order to connect any two refineries. The pipeline diameter considered in all cases was 8 inches.

The petrochemical complex for the production of PVC starts with the production of ethylene from the refineries feedstocks. The main feedstocks to the ethylene plant in our study are light naphtha (LN) and gas oil (GO). The selection of the feedstocks and, hence, the process technologies is based on the optimal balance and trade-off between the refinery and petrochemical markets. The process technologies considered in this study for the production of PVC are list in Table 5.2. The overall topology of all petrochemical technologies for PVC production is shown in Figure 5.3.

From the refinery side, the proposed model will provide the optimal network integration between the refineries, process expansion requirements, operating policy based on different feedstock combination alternatives, process levels and operating modes. On the petrochemical side, the model will establish the design of an optimal petrochemical process network for the production of PVC from the range of process technologies and feedstocks available to satisfy a given demand.

Table 5.2 Major products and process technologies in the petrochemical complex.

Product	Sale price ($/ton)[a]	Process technology	Process index	Min econ. prod. (10^3 t/year)
Ethylene (E)	1570	Pyrolysis of naphtha (low severity)	1	250
		Pyrolysis of gas oil (low severity)	2	250
		Steam cracking of naphtha (high severity)	3	250
		Steam cracking of gas oil (high severity)	4	250
Ethylene dichloride (EDC)	378	Chlorination of ethylene	5	180
		Oxychlorination of ethylene	6	180
Vinyl chloride monomer (VCM)	1230	Chlorination and Oxychlorination of ethylene	7	250
		Dehydrochlorination of ethylene dichloride	8	125
Polyvinyl chloride (PVC)	1600	Bulk polymerization	9	50
		Suspension polymerization	10	90

a) All chemical prices in this study were obtained from latest CW Price Reports in the Chemical Week journal.

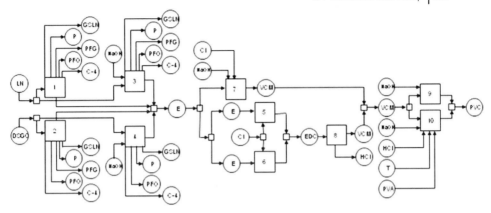

Figure 5.3 SEN representation of the PVC petrochemical complex possible routes.

This problem was formulated as an MILP with the overall objective of minimizing total annualized cost of the refinery and maximizing the added value from the PVC petrochemical network. Maximizing the added value of the petrochemical network is appropriate since the feedstock costs contribute to the majority of the total cost. For instance, the feedstock cost of an ethylene plant contributes to more than 87% of the total cost when naphtha is used and 84% and 74% when propane and ethane are used, respectively (NBK MENA Equity Research, 2007).

The modeling system GAMS (Brooke et al., 1996) is used for setting up the optimization models. The computational tests were carried out on a Pentium M processor 2.13 GHz and the MILP problems were solved with CPLEX (CPLEX Optimization, Inc., 1993).

The problem was first solved for the refinery network separately in order to compare and illustrate the effect of considering the PVC complex on the refinery network design and operating policies. Table 5.3 shows the optimal network integration design and operating policies of the refineries. The three refineries collaborated to satisfy a given local market demand and the model proposed the production and blending level targets for the individual sites. The annual production cost across the facilities was found to be $9 331 000.

The model was then solved for the total refinery network and the PVC petrochemical complex. As shown in Table 5.4, the proposed model redesigned the refinery network and operating policies and also devised the optimal production plan for the PVC complex from all available process technologies. The model selected gas oil, an intermediate refinery stream, as the refinery feedstock to the petrochemical complex as opposed to the normally used light naphtha feedstock in industrial practice. In fact, this selection provided the optimal strategy as the light naphtha stream was used instead in the gasoline

Table 5.3 Model results of multi-refinery network.

Process variables				Results (10^3 t/year)		
				R1	R2	R3
Crude oil supply				4500.0	12 000.0	9900.0
Production levels	Crude unit			4500.0	12 000.0	9900.0
	Reformer			573.1	1824.5	1800.0
	Isomerization			200.0	—	450.0
	FCC			640.0	1400.0	—
	Hydrocracker			—	1740.4	2400
	Delayed coker			—	—	1484
	Des gas oil			1084.6	2763.7	2383.8
	Des cycle gas oil			200.0	600.0	—
	Des ATK			—	1200.0	1654.8
	Des distillates			—	—	360
Intermediate streams exchange	From	R1	VGO	—	—	446.0 to HCU
			VRSD	—	—	380.4 to Coker
		R2	LN	—	—	340.0 to Isomer
		R3	LN	86.1 to Isomer	—	—
			VGO	—	144.7 to FCC	—
			UCO	—	130.1 to FCC	—
Exports	PG95				273.2	
	JP4				1359.5	
	GO6				3695.7	
	HFO				1579.8	
	ATK				1767.5	
	Coke				203.9	
	Diesel				109.7	
Total cost ($/yr)					9 331 000	

pool for maximum gasoline production. PVC production was proposed by first the high severity steam cracking of gas oil to produce ethylene. Vinyl chloride monomer (VCM) is then produced through the chlorination and oxychlorination of ethylene and, finally, VCM is converted to PVC by bulk polymerization. The simultaneous optimization of the refinery and petrochemical network had an impact on both the refinery intermediate exchange network and production levels, as shown in Tables 5.3 and 5.4. For example, the capacity utilization of the desulfurization of gas oil process increased from 83%, 92%, 99% in refineries 1, 2, and 3, respectively, to 100% in all refineries when the petrochemical network was considered in the model. The annual production cost across the facilities was found to be $8 948 000.

Table 5.4 Deterministic model results of refinery and petrochemical networks.

	Process variables			Results (10³ t/year)		
				R1	R2	R3
Refinery	Crude oil supply			4500.0	12 000.0	9900.0
	Production levels	Crude unit		4500.0	12 000.0	9900.0
		Reformer		573.1	1824.6	1793.5
		Isomerization		200.0	—	450.0
		FCC		640.0	1400.0	—
		Hydrocracker		—	1740.4	2400.0
		Delayed coker		—	—	1440.0
		Des gas oil		1300.0	3000.0	2400.0
		Des cycle gas oil		200.0	600.0	—
		Des ATK		—	1200.0	1654.8
		Des distillates		—	—	360.0
	Intermediate streams exchange	From	R1 VGO	—	204.4 to HCU	301.2 to HCU
			R2 LN	—	—	321.2 to Isom
			VRSD	—	—	267.6 to Coker
			R3 LN	68.0 to Isom		
			UCO	—	100.3 to FCC	—
	Exports	PG95			265.2	
		JP4			1365.5	
		GO6			1503.4	
		HFO			1658.9	
		ATK			1767.5	
		Coke			178.8	
		Diesel			84.0	
Petrochemical	Refinery feed to PVC complex	Gas oil		1162.4	920.0	71.3
	Production levels	S. Crack GO (4)			552.2	
		Cl and OxyCl E (7)			459.1	
		Bulk polym. (9)			204.0	
	Final Products	PVC			204.0	
Total cost ($/yr)					$8 948 000	

5.5 Conclusion

A mixed-integer programming model for designing an integration and coordination policy among multiple refineries and a petrochemical network was presented.

The objective was to develop a simultaneous methodology for designing a process integration network between petroleum refining and the petrochemical industry. A large-scale three refinery network and a PVC petrochemical complex were integrated to illustrate the performance of the proposed design methodology and to show the economic potential and trade-offs involved in the optimization of such systems. The study showed that the optimization of the downstream petrochemical industry has an impact on the multi-refinery network integration and coordination strategies. This result emphasizes the importance of the developed methodology.

In this Chapter, however, all parameters were assumed to be known with certainty. Nevertheless, the current situation of fluctuating and high petroleum crude oil prices, changes in demand, and the direct effect this can have on the downstream petrochemical system underlines the importance of considering uncertainties. For example, the availability of crude oils, feedstock and chemicals prices, and market demands for finished products will have a direct impact on the output of the highly strategic decisions involved in our study. Acknowledging the shortcomings of deterministic models, the next part of this book will consider uncertainties in the integration problem.

References

Brooke, A., Kendrick, D., Meeraus, A., and Raman, R. (1996) *GAMS–A User's Guide*, GAMS Development Corporation, Washington DC.

CPLEX Optimization, Inc. (1993) Using the CPLEX Callable Library and CPLEX Mixed Integer Library, CPLEX Optimization, Inc., Incline Village, NV.

Gonzalo, M.F., Balseyro, I.G., Bonnardot, J., Morel, F., and Sarrazin, P. (2004) Consider integrating refining and petrochemical operations. *Hydrocarbon Processing*, **83**, 61.

Joly, M., Moro, L.F.L., and Pinto, J.M. (2002) Planning and scheduling for petroleum refineries using mathematical programming. *Brazilian Journal of Chemical Engineering*, **19**, 207.

Kuo, T., and Chang, T. (2008) Optimal planning strategy for the supply chains of light aromatic compounds in petrochemical industries. *Computers & Chemical Engineering*, **32**, 1147.

Li, C., He, X., Chen, B., Chen, B., Gong, Z., and Quan, L. (2006) Integrative optimization of refining and petrochemical plants. Proceedings of the 16th European Symposium on Computer Aided Process Engineering and 9th International Symposium on Process Systems Engineering; Garmisch-Partenkirchen, Germany, July 9–13, p. 21.

Lippe, D. (2007) Gasoline price spike shifts ethylene feeds. *Oil & Gas Journal*, **105**, 56.

Lippe, D. (2008) US olefins see improving feed economics, demand. *Oil & Gas Journal*, **106**, 56.

NBK MENA Equity Research (2007) Petrochemical Primer. http://cm2.zawya.com/researchreports/nbk/20070923_NBK_105729.pdf, accessed on March 20, 2008.

Rudd, D.F., Fathi-Afshar, S., Trevino, A.A., and Statherr, M.A. (1981) *Petrochemical Technology Assessment*, John Wiley & Sons, Inc., New York.

Sadhukhan, J., Zhang, N., and Zhu, X.X. (2004) Analytical optimization of industrial systems and applications to refineries, petrochemicals. *Chemical Engineering Science*, **59**, 4169.

Sahinidis, N.V., Grossmann, I.E., Fornari, R.E., and Chathrathi, M. (1989) Optimization model for long range planning in chemical industry. *Computers & Chemical Engineering*, **13**, 1049.

Shah, N. (1998) Single- and multisite planning and scheduling: current status and future challenges. AIChE Symposium Series: *Proceedings of the Third International Conference of the Foundations of Computer-Aided Process Operations, Snowbird, Utah, USA, July 5–10*, American Institute of Chemical Engineering, **94**, p. 75.

Singh, J., Kumar, M.M., Saxena, A.K., and Kumar, S. (2005) Reaction pathways and product yields in mild thermal cracking of vacuum residues: A multi-lump kinetic model. *Chemical Engineering Journal*, **108**, 239.

Swaty, T.E. (2002) Consider over-the-fence product stream swapping to raise profitability. *Hydrocarbon Processing*, **81**, 37.

Van Geem, K.M., Reyniers, M.F., and Marin, G.B. (2008) Challenges of modeling steam cracking of heavy feedstocks. *Oil & Gas Science & Technology*, **63**, 79.

Part Three
Planning Under Uncertainty

6
Planning Under Uncertainty for a Single Refinery Plant

In this chapter, we study petroleum refining problems under uncertainty through a simplified example in order to explain the concept of stochastic programming. In particular, we concentrate on two-stage stochastic programming and explain the physical meaning of recourse concept and how it applies to the petroleum industry. We also explain the idea of risk management through different mathematical representations and provide detailed analysis of two main risk measures; namely, (i) variance and (ii) mean-absolute deviation. The different models are numerically explained using the refinery LP model introduced in Chapter 2.

6.1
Introduction

Chemical process systems are subject to uncertainties due to many random events such as raw material variations, demand fluctuations, equipment failures, and so on. In this chapter we will utilize stochastic programming (SP) methods to deal with these uncertainties that are typically employed in computational finance applications. These methods have been very useful in screening alternatives on the basis of the expected value of economic criteria as well as the economic and operational risks involved. Several approaches have been reported in the literature addressing the problem of production planning under uncertainty. Extensive reviews surveying various issues in this area can be found in Applequist et al. (1997), Shah (1998), Cheng, Subrahmanian, and Westerberg (2005) and Méndez et al. (2006).

Problems of design and planning of chemical processes and plants under uncertainty have been effectively addressed in the process systems engineering (PSE) literature using the two-stage stochastic programming (SP) with recourse model. Under this framework, the problem is posed as one of optimizing an objective function that consists of two stages. The first corresponds to decisions on the global or planning variables, whose fixed values are selected ahead of the realization of the uncertain events. The second term represents the expected value of the decisions due to the production variables, whose flexible values will be adjusted to achieve feasibility during operation (Acevedo and Pistikopoulos, 1998).

Planning and Integration of Refinery and Petrochemical Operations. Khalid Y. Al-Qahtani and Ali Elkamel
Copyright © 2010 WILEY-VCH Verlag GmbH & Co. KGaA, Weinheim
ISBN: 978-3-527-32694-5

The presence of uncertainty is translated into the stochastic nature of the recourse costs associated with the second-stage decisions. Hence, the goal in the two-stage modeling approach to planning decisions under uncertainty is to commit initially to the planning variables in such a way that the sum of the first-stage costs and the expected value plus deviations of the typically more expensive random second-stage recourse costs is minimized (Swamy and Shmoys, 2006). Approaches differ primarily in the way the expected value and its deviation terms are evaluated.

6.2
Problem Definition

The midterm refinery production planning problem addressed in this chapter can be stated as follows. It is assumed that the physical resources of the plant are fixed and that the associated prices, costs, and demands are externally imposed. The objective is to determine the optimal production plan by computing the amount of materials that are processed at each time in each unit, in the face of three major uncertainty sources. The uncertainty is considered in terms of (i) market demand for products; (ii) prices of crude oil and products; and (iii) product yields of crude oil from the crude distillation unit. A hybrid stochastic programming technique is applied within a framework of the classical two-stage stochastic programming with recourse. The risk analysis in our problem follows the mean-variance (E-V or MV) portfolio optimization formulation of Markowitz (1952) in both profit and the recourse penalty costs. A numerical study based on the deterministic refinery planning model of Ravi and Reddy (1998) and Allen (1971) is utilized to demonstrate the implementation of the proposed approaches without loss of generality. The single-objective linear programming (LP) model is first solved deterministically and is then reformulated with the addition of the stochastic dimension according to principles and approaches outlined under the general model development below.

6.3
Deterministic Model Formulation

The basic framework for the deterministic planning model is based on the formulation presented by Ierapetritou and Pistikopoulos (1994). Consider the production planning problem of a typical refinery operation with a network of M continuous processes and N materials, as shown in Figure 6.1. Let $j \in J$ index the set of continuous processes whereas $i \in I$ indexes the set of materials. These products are produced during n time periods indexed by $t \in T$ to meet a prespecified level of demand during each period. Given also are the prices and availabilities of materials as well as investment and operating cost data over a time period. A typical aggregated mixed-integer linear planning model consists of the following sets of constraints and objective function.

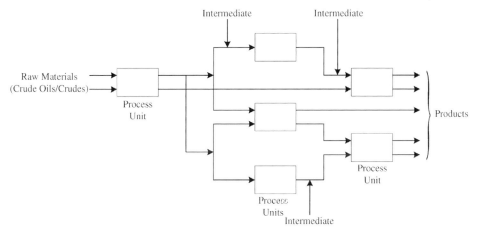

Figure 6.1 General block flow diagram of a refinery/chemical plant.

1) **Production capacity constraints:**

$$x_{j,t} = x_{j,t-1} + CE_{j,t} \quad \forall j \in J \tag{6.1}$$

$$y_{j,t} CE_{j,t}^L \leq CE_{j,t} \leq y_{j,t} CE_{j,t}^U \quad \forall j \in J, \forall t \in T \tag{6.2}$$

where

$$y_{j,t} = \begin{bmatrix} 1 & \text{if there is an expansion} \\ 0 & \text{otherwise} \end{bmatrix} \tag{6.3}$$

2) **Demand constraints:**

$$S_{i,t} + L_{i,t} = d_{i,t}, \quad \forall i \in I', \forall t \in T \tag{6.4}$$

$$d_{i,t}^L \leq S_{i,t} \leq d_{i,t}^U, \quad \forall i \in I', \forall t \in T \tag{6.5}$$

3) **Availability constraints:**

$$p_t^L \leq P_t \leq p_t^U, \quad \forall i \in I', \forall t \in T \tag{6.6}$$

4) **Inventory requirements:**

$$I_{i,t}^{f\,min} \leq I_{i,t}^f \leq I_{i,t}^{f\,max} \quad \forall i \in I' \tag{6.7}$$

$$I_{i,t}^f = I_{i,t+1}^s, \quad \forall i \in I', \forall t \in T \tag{6.8}$$

This constraint is needed to indicate that a certain level of inventory must be maintained at all times to ensure material availability, in addition to the amount of materials purchased and/or produced. Equation 6.8 simply states that $I_{i,t+1}^s$, the starting inventory of material i in period $t+1$ is the same as $I_{i,t}^f$, the inventory of material i at the end of the preceding period t (if $t=1$, then $I_{i,t}^f = I_{i,1}^f$ denotes the initial inventory).

5) **Material balances**:

$$P_t + I_{i,t}^s + \sum_{j \in J} b_{i,j} x_{j,t} - S_{i,t} - I_{i,t}^f = 0, \qquad \forall i \in I', \forall t \in T \tag{6.9}$$

6) **Objective function**: a profit maximization function over the time horizon is considered as the difference between the revenue due to product sales and the overall costs, with the latter consisting of the cost of raw materials, operating cost, investment cost, and inventory cost:

$$\max \text{ profit } z_0 = \sum_{t \in T} \begin{bmatrix} \sum_{i \in I} \gamma_{i,t} S_{i,t} + \sum_{i \in I} \tilde{\gamma}_{i,t} I_{i,t}^f - \sum_{i \in I} \lambda_{i,t} P_{i,t} - \sum_{i \in I} \tilde{\lambda}_{i,t} I_{i,t}^s - \sum_{j \in J} C_{j,t} x_{j,t} \\ - \sum_{i \in I} h_{i,t} H_{i,t} - \sum_{j \in J} (\alpha_{j,t} CE_{j,t} + \beta_{j,t} Y_{j,t}) - (r_t R_t + o_t O_t) \end{bmatrix}$$

(6.10)

6.4
Stochastic Model Formulation

In spite of the resulting exponential increase in the problem size with the number of uncertain parameters, the scenario analysis approach has been used considerably in the literature and has been proven to provide reliable and practical results for optimization under uncertainty (Gupta and Maranas, 2003). Hence, in this chapter, it is adopted for describing uncertainty in the stochastic parameters. Representative scenarios are constructed to model uncertainty in the random variables of prices, demands, and yields within the two-stage stochastic programming (SP) framework. We will refrain at this stage from explaning methods to quantify the required number of scenarios to achieve a given confidence interval. This subject will be explained thourougly in Chapter 7.

6.4.1
Appraoch 1: Risk Model I

The first approach adopts the classical Markowitz's MV model to handle randomness in the objective function coefficients of prices, in which the expected profit is maximized while an appended term representing the magnitude of operational risk due to variability or dispersion in price, as measured by variance, is minimized (Eppen, Martin, and Schrage, 1989). The model can be formulated as minimizing risk (i.e., variance) subject to a lower bound constraint on the target profit (i.e., the mean return).

Malcolm and Zenios (1994) presented an application of the MV approach by adopting the robust optimization framework proposed by Mulvey, Vanderbei, and

Zenios (1995) to the problem of capacity expansion of power systems. The problem was formulated as a large-scale nonlinear program with variance of the scenario-dependent costs included in the objective function. Another application using variance is employed by Bok, Lee, and Park (1998), also within a robust optimization framework of Mulvey, Vanderbei, and Zenios (1995), for investment in the long-range capacity expansion of chemical process networks under uncertain demands.

6.4.1.1 Sampling Methodolgy

A collection of scenarios is generated that best captures the trend of raw material prices of the different types of crude oil and the sales prices of the saleable refining products for a representative period of time based on available historical data. A probability p_s, with index s denoting the sth scenario, is assigned to each scenario to reflect the likelihood of each scenario being realized with $\sum_{s \in S} p_s = 1$.

6.4.1.2 Objective Function Evaluation

To represent the different scenarios accounting for uncertainty in prices, the price-related random objective function coefficients comprising: (i) $\lambda_{i,t}$ for the costs of different types of crude oil that can be handled by the crude distillation unit of a refinery and (ii) $\gamma_{i,t}$ for the sales prices of the refined products, are added with an index s subscript, each with an associated probability p_s. For ease of reference, both groups of price (or cost) parameters are redefined as the parameter $c_{i,s,t}$ or $c_{i_{cr},s,t}$; the difference between the two is in the use of the index i_{cr} (and the corresponding set of I_{cr}) to refer to "products" that are actually crude oils, as distinguished from the index i that is used to indicate the saleable products. Since the objective function given by Equation 6.10 is linear, it is straightforward to show that the expectation of the random objective function with random price coefficients is given by:

$$E[z_0] = \sum_{t \in T} \left[\sum_{i \in I} \sum_{s \in S} p_s c_{i,s,t} S_{i,t} + \sum_{i \in I} \tilde{\gamma}_{i,t} I_{i,t}^f - \sum_{i' \in I_{cr}} \sum_{s \in S} p_s c_{i_{cr},s,t} P_t - \sum_{i \in I} \tilde{\lambda}_{i,t} I_{i,t}^s - \sum_{i \in I} h_{i,t} H_{i,t} \right. \\ \left. - \sum_{j \in J} (\alpha_{j,t} CE_{j,t} + \beta_{j,t} Y_{j,t}) + r_t R_t + o_t O_t \right]$$

(6.11)

Consideration of the expected value of profit alone as the objective function, which is characteristic of the classical stochastic linear programs introduced by Dantzig (1955) and Beale (1955), is obviously inappropriate for moderate and high-risk decisions under uncertainty since most decision makers are risk averse in facing important decisions. The expected value objective ignores both the risk attribute of the decision maker and the distribution of the objective values. Hence, variance of each of the random price coefficients can be adopted as a viable risk measure of the objective function, which is the second major component of the MV approach adopted in Risk Model I.

6.4.1.3 Variance Calculation

Variance for the expected value of the objective function (6.10) is expressed as:

$$V(z_0) = \sum_{t \in T} \sum_{i \in I} S_{i,t}^2 V(c_{i,s,t}) + \sum_{t \in T} \sum_{i' \in I'} P_{i',t}^2 V(c_{i',s,t}) \tag{6.12}$$

Since the above derivation does not explicitly evaluate variances of the random price coefficients as given by $V(c_{i,s,t})$ and $V(c_{i',s,t})$, we consider the following alternative definition for variance from Markowitz (1952) that yields:

$$V(z_0) = p_{s_1}(z_{s_1} - E[z_0])^2 + p_{s_2}(z_{s_2} - E[z_0])^2 + \cdots + p_{s_\omega}(z_{s_\omega} - E[z_0])^2 \tag{6.13}$$

The objective function for the stochastic model is now given by:

$$\max z_1 = E[z_0] - \theta_1 V(z_0)$$
$$\text{subject to constraints } (6.1) \text{ to } (6.8) \tag{6.14}$$

The model is subject to the same set of constraints as the deterministic model, with θ_1 as the risk trade-off parameter (or simply termed the risk factor) associated with risk reduction for the expected profit. θ_1 is varied over the entire range of $(0, \infty)$ to generate a set of feasible decisions that have maximum return for a given level of risk, which is equivalent to the "efficient frontier" portfolios for investment applications.

It is noteworthy that from a modeling perspective, θ_1 is also a scaling factor, since the expectation operator and the variance are of different dimensions. If it is desirable to obtain a term that is dimensionally consistent with the expected value term, then the standard deviation of z_0 may be considered, instead of the variance, as the risk measure (in which standard deviation is simply the square root of variance). Moreover, θ_1 represents the weight or weighting factor for the variance term in a multiobjective optimization setting that consists of the components' mean and variance.

However, the primary difficulty in executing model (6.14) is in determining a suitable set of values for θ_1 that caters to decision makers who are either risk-prone or risk-averse. An approach to circumvent this problem is available in which the variance (or the standard deviation) of the objective function is minimized as follows:

$$\max z_1 = -V(z_0)$$
$$(\text{or } \max z_1 = -\sqrt{V(z_0)}) \tag{6.15}$$

while adding the inequality constraint for the mean of the objective function that stipulates a certain target value for the desired profit to be achieved:

$$E[z_0] \geq \text{Target profit} \tag{6.16}$$

Thus, the final form of Risk Model I is given by:

$$\max z_1 = -V(z_0)$$
$$\text{subject to } E[z_0] \geq \text{Target profit} \quad \text{(RM1)}$$
$$\text{Constraints } (6.1) - (6.10)$$

To determine a suitable range for the target profit value, a test value is assumed and the corresponding solution is computed. Then, the test value is increased or decreased, with the solution computed each time to investigate and establish the range of target values that ensures solution feasibility.

6.4.2
Approach 2: Expectation Model I and II

In Approach 2, the MV model developed in Approach 1 is incorporated within a two-stage SP with fixed recourse framework to handle randomness in the right-hand side (RHS) and left-hand side (LHS) coefficients of the related constraints.

6.4.2.1 Demand Uncertainty

Uncertainty in market demand introduces randomness in constraints for production requirements of intermediates and saleable products, as given by Equation (6.4). The sampling methodology employed for scenario construction is similar to the case of price uncertainty in Approach 1, involving the generation of representative scenarios of demand uncertainty for N number of products with the associated probabilities that indicate their comparative frequency of occurrence.

One of the main consequences of uncertainty within the context of decision-making is the possibility of infeasibility in the future. The two-stage recourse modeling framework provides the liberty of addressing this issue by postponing some decisions into the second stage; however, this comes at the expense of the use of corresponding penalties in the objective function. Decisions that can be delayed until after information about the uncertain data is available almost definitely offer an opportunity to adjust and adapt to the new information received. There is generally value associated with delaying a decision, when it is possible to do so, until after additional information is obtained.

In devising the appropriate penalty functions, we resort to the introduction of compensating slack variables in the probabilistic constraints to eliminate the possibility of second-stage infeasibility. Additionally, the recourse-based modeling philosophy requires the decision maker to assign a price as a penalty to remedial activities that are undertaken in response to uncertainty. For applications in production planning, these can be assumed as fixed standard costs. However, under some circumstances, it may be more appropriate to accept the possibility of infeasibility, provided that the probability of this event is restricted below a given threshold. This is addressed in the subsequent approaches by appending an appropriate risk measure to the objective function.

Compensating slack variables accounting for shortfall and/or surplus in production are introduced in the stochastic constraints with the following results: (i) inequality constraints are replaced with equality constraints; (ii) numerical feasibility of the stochastic constraints can be ensured for all events; and (iii) penalties for feasibility violations can be added to the objective function. Since a probability can be assigned to each realization of the stochastic parameter vector (i.e., to each scenario), the probability of feasible operation can be measured. In this work, a non-negative second-stage recourse slack variable $z_{i,s}^+$ quantifies the shortfall in production, which is penalized in the objective function according to the cost of purchasing this make-up product from the open market. Likewise, for overproduction (surplus) with respect to market demands, the recourse slack variable $z_{i,s}^-$ is penalized based on the inventory cost for storing the excess of production. The expected values of the recourse penalty costs of c_i^+ and c_i^- for infeasibility due to shortfall and surplus of production, respectively, are minimized in the objective function in an effort to maximize the expected profit. Thus, the expected recourse penalty for the second-stage costs due to uncertainty in the demand for product i for all considered scenarios is given by:

$$E_{s,\text{demand}} = \sum_{i \in I} \sum_{s \in S} p_s (c_i^+ z_{i,s}^+ + c_i^- z_{i,s}^-) \qquad (6.17)$$

To ensure that the original information structure associated with the decision process sequence is honored, for each of the products whose demand is uncertain, the number of new constraints to be added to the stochastic model counterpart, replacing the original deterministic constraint, corresponds to the number of scenarios. Herein lies a demonstration of the fact that the size of a recourse model increases exponentially since the total number of scenarios grows exponentially with the number of random parameters. In general, the new constraints take the form:

$$S_{i,t} + z_{i,s}^+ - z_{i,s}^- = d_{i,t,s}, \qquad \forall i \in I_P, \forall t \in T, \forall s \in S \qquad (6.18)$$

6.4.2.2 Process Yield Uncertainty

Uncertainty in product yields introduces randomness in the material balances that are given by Equation (6.9). The scenario construction to model yield uncertainty of products k from material i is similar to the approach for modeling demand uncertainty. Note that in order to ensure that the material balances are satisfied, the summation of yields must be equal to unity. The non-negative second-stage recourse slack variables $y_{i,k,s}^+$ and $y_{i,k,s}^-$ represent shortage and excess in yields, respectively, with their corresponding fixed unit recourse penalty costs given by $q_{i,k}^+$ and $q_{i,k}^-$. Thus, the expected recourse penalty for the second-stage costs due to yield uncertainty is:

$$E_{s,\text{yield}} = \sum_{i \in I} \sum_{s \in S} \sum_{k \in K} p_s (q_{i,k}^+ y_{i,k,s}^+ + q_{i,k}^- y_{i,k,s}^-) \qquad (6.19)$$

N_s new constraints to represent the N_s number of scenarios dealing with yield uncertainty are introduced for each product whose yield is uncertain, with the general form of the new constraints given by:

$$P_t + I_{i,t}^s + \sum_{j \in J} b_{i,j} x_{j,t} + y_{i,k,s}^+ - y_{i,k,s}^- - S_{i,t} - I_{i,t}^f = 0, \qquad i \in I, k \in K, s \in S \qquad (6.20)$$

The two major assumptions that enable the combination of the sub-scenarios are that: (i) the uncertain parameters of prices, demands, and yields in each scenario are highly-correlated; and (ii) each of the random variables (or equivalently, each of the scenarios) is assumed to be independent of any another. These assumptions lead to two implications: (i) they obviate the need to construct a joint probability distribution function (in the sampling methodology) that encompasses scenarios of all the possible combinations of the three random variables (this means that, for instance, the possibility of a scenario in which prices are "average" with demand being "above average" and yield being "below average" is not considered); (ii) the covariance term in the MV model becomes equal to variance (Bernardo 1999).

The corresponding expected recourse penalty for the second-stage costs is given by:

$$E_s = E_{s,\text{demand}} + E_{s,\text{yield}} = \sum_{i \in I} \sum_{s \in S} p_s [(c_i^+ z_{i,s}^+ + c_i^- z_{i,s}^-) + (q_i^+ y_{i,s}^+ + q_i^- y_{i,s}^-)] = \sum_{i \in I} \sum_{s \in S} p_s \xi_s \quad (6.21)$$

where $\xi_{i,s} = (c_i^+ z_{i,s}^+ + c_i^- z_{i,s}^-) + (q_i^+ y_{i,k,s}^+ + q_i^- y_{i,k,s}^-)$. Thus, Expectation Model I is formulated as:

$$\max z_2 = z_1 - E_s = [z_0] - \theta_1 V(z_0) - E_s$$
subject to constraints (6.1) – (6.3), (6.6) – (6.8), (6.18) and (6.20) \quad (EM1)

As remarked in Approach 1, a potential complication with Expectation Model I lies in computing a suitable range of values for the operational risk factor θ_1. Therefore, an alternative formulation of minimizing variance while adding a target profit constraint is employed for Expectation Model II:

$$\max z_2 = -V(z_0) - E_s$$
subject to $E[z_0] \geq$ Target profit \quad (EM2)
constraints (6.1) – (6.3), (6.6) – (6.8), (6.18) and (6.20)

6.4.3
Approach 3: Risk Model II

The goal of Approach 3 is to append an operational risk term to the mean-risk model formulation in Approach 2 to account for the significance of both financial risk (as considered by Approach 1) and operational risk in decision-making.

Variance for the various expected recourse penalty for the second-stage costs V_s is derived as:

$$V_s = \sum_{s \in S} p_s (\xi_s - E_s)^2 = \sum_{s \in S} p_s \left(\xi_s - \sum_{s' \in S'} p_{s'} \xi_{s'} \right)^2$$

$$\Rightarrow V_s = \sum_{s \in S} p_s \left\{ \sum_{i \in I} \begin{bmatrix} (c_i^+ z_{i,s}^+ + c_i^- z_{i,s}^-) \\ + (q_i^+ y_{i,k,s}^+ + q_i^- y_{i,k,s}^-) \end{bmatrix} \right. \quad (6.22)$$

$$\left. - \sum_{i \in I} \sum_{s' \in S'} p_{s'} \begin{bmatrix} (c_i^+ z_{i,s'}^+ + c_i^- z_{i,s'}^-) \\ + (q_i^+ y_{i,k,s'}^+ + q_i^- y_{i,k,s'}^-) \end{bmatrix} \right\}^2$$

Note that the index s' and the corresponding set S' is used to denote scenarios for the evaluation of the inner expectation term to distinguish them from the original index s used to represent the scenarios. V_s is weighted by the operational risk factor $\theta_2 \in (0, \infty)$. The formulation of Risk Model II is:

$$\max z_3 = z_2 - \theta_2 V_s = E[z_0] - \theta_1 V(z_0) - E_{s'} - \theta_2 V_s \quad \text{(RM2)}$$
$$\text{Constraints } (6.1) - (6.3), (6.7) - (6.9), (6.24) \text{ and } (6.26)$$

6.4.4
Approach 4: Risk Model III

Konno and Yamazaki (1991) proposed a large-scale portfolio optimization model based on mean-absolute deviation (MAD). This serves as an alternative measure of risk to the standard Markowitz's MV approach, which models risk by the variance of the rate of return of a portfolio, leading to a nonlinear convex quadratic programming (QP) problem. Although both measures are almost equivalent from a mathematical point-of-view, they are substantially different computationally in a few perspectives, as highlighted by Konno and Wijayanayake (2002) and Konno and Koshizuka (2005). In practice, MAD is used due to its computationally-attractive linear property.

Therefore, in this approach, we develop Risk Model III as a reformulation of Risk Model II by employing the mean-absolute deviation (MAD), in place of variance, as the measure of operational risk imposed by the recourse costs to handle the same three factors of uncertainty (prices, demands, and yields). To the best of our knowledge, this is the *first* such application of MAD, a widely-used metric in the area of system identification and process control, for risk management in refinery planning.

The L_1 risk of the absolute deviation function is given by Konno and Yamazaki (1991):

$$W(\mathbf{x}) = E\left[\left|\sum_{j=1}^n R_j x_j - E\left[\sum_{j=1}^n R_j x_j\right]\right|\right] \quad (6.23)$$

Thus, the corresponding mean-absolute deviation (MAD) of the expected penalty costs is formulated as:

$$W = \sum_{s \in S} p_s |\xi_s - E_{s'}| = \sum_{s \in S} p_s \left|\xi_s - \sum_{s' \in S'} p_{s'} \xi_{s'}\right|$$
$$\Rightarrow W = \sum_{s \in S} p_s \left|\sum_{i \in I} \begin{bmatrix} (c_i^+ z_{i,s}^+ + c_i^- z_{i,s}^-) \\ + (q_i^+ y_{i,s}^+ + q_i^- y_{i,s}^-) \end{bmatrix} - \sum_{i \in I} \sum_{s' \in S'} p_{s'} \begin{bmatrix} (c_i^+ z_{i,s'}^+ + c_i^- z_{i,s'}^-) \\ + (q_i^+ y_{i,k,s'}^+ + q_i^- y_{i,k,s'}^-) \end{bmatrix}\right| \quad (6.24)$$

This nonlinear function can be linearized by implementing the transformation procedure outlined by Papahristodoulou and Dotzauer (2004), in which W must satisfy the following conditions:

$$W \geq -\sum_{s \in S} p_s \left\{ \sum_{i \in I} \left[\begin{array}{c} (c_i^+ z_{i,s}^+ + c_i^- z_{i,s}^-) \\ + (q_i^+ y_{i,s}^+ + q_i^- y_{i,s}^-) \end{array} \right] - \sum_{i \in I} \sum_{s' \in S'} p_{s'} \left[\begin{array}{c} (c_i^+ z_{i,s'}^+ + c_i^- z_{i,s'}^-) \\ + (q_i^+ y_{i,k,s'}^+ + q_i^- y_{i,k,s'}^-) \end{array} \right] \right\} \quad (6.25)$$

$$W \geq \sum_{s \in S} p_s \left\{ \sum_{i \in I} \left[\begin{array}{c} (c_i^+ z_{i,s}^+ + c_i^- z_{i,s}^-) \\ + (q_i^+ y_{i,s}^+ + q_i^- y_{i,s}^-) \end{array} \right] - \sum_{i \in I} \sum_{s' \in S'} p_{s'} \left[\begin{array}{c} (c_i^+ z_{i,s'}^+ + c_i^- z_{i,s'}^-) \\ + (q_i^+ y_{i,k,s'}^+ + q_i^- y_{i,k,s'}^-) \end{array} \right] \right\} \quad (6.26)$$

$$W \geq 0 \quad (6.27)$$

Similar to Risk Model II, the adoption of MAD is weighted by the operational risk factor θ_3 ($0 < \theta_3 < \infty$) in Risk Model III, to give the following formulation:

$\max z_4 = z_2 - \theta_3 W = E[z_0] - \theta_1 V(z_0) - E_{s'} - \theta_3 W$

Constraints (6.1) – (6.3), (6.6) – (6.8), (6.18) and (6.20)   (RM3)

MAD linearization conditions (6.25) – (6.27)

6.5 Analysis Methodology

In the context of production planning, robustness can generally be defined as a measure of the resilience of the planning model to respond in the face of parameter uncertainty and unplanned disruptive events, (Vin and Ierapetritou, 2001). To investigate and interpret the behavior and overall robustness of the proposed multiobjective optimization models in this chapter, we carry out a series of rigorous computational experiments to establish the effectiveness of the stochastic models in hedging against uncertainties posed by randomness in prices, demands, and yields. Two performance metrics that have been previously utilized in the optimization literature are considered to quantitatively measure and account for characteristics of planning under simultaneous uncertainty in prices, demands, and yields. The two metrics are: (i) the concepts of solution robustness and model robustness, and (ii) the coefficient of variation C_v.

6.5.1 Model and Solution Robustness

It is desirable to demonstrate that the proposed stochastic formulations provide robust results. According to Mulvey, Vanderbei, and Zenios (1995), a robust solution remains close to optimality for all scenarios of the input data while a robust model remains almost feasible for all the data of the scenarios. In refinery planning, model robustness or model feasibility is as essential as solution optimality. For example, in mitigating demand uncertainty, model feasibility is represented by an optimal solution that has almost no shortfalls or surpluses in production. A trade-off exists between solution optimality and model-and-solution robustness. To investigate these trends, the following parameters are analyzed from the raw computational results of the refinery production rates for the models:

- the expected deviation in profit as measured by variance $V(z_0)$;
- the expected total unmet demand (i.e., production shortfall);
- the expected total excess production (i.e., production surplus);and the expected recourse penalty costs E_s.

6.5.2
Variation Coefficient

To interpret the solutions obtained from the stochastic models, we propose to investigate their corresponding coefficient of variation C_v. C_v for a set of values is defined as the ratio of the standard deviation to the expected value or mean and is usually expressed as a percentage. It is calculated as:

$$C_v = \frac{\text{standard deviation}}{\text{mean}} \times 100\% = \frac{\sigma}{\mu} \times 100\% = \frac{\sqrt{V}}{E} \times 100\% \qquad (6.28)$$

Statistically, C_v is a measure of reliability, or evaluated from the opposite but equivalent perspective, it is also indicative of the degree of uncertainty. It is alternatively interpreted as the inverse ratio of data to noise in the data in signal-processing-related applications. Thus, it is apparent that a small value of C_v is desirable as it signifies a small degree of noise or variability (e.g., in a data set) and, hence, reflects low uncertainty.

In stochastic optimization, C_v can be purposefully employed to investigate, denote, and compare the relative uncertainty in models being studied. In a risk minimization model, as the expected value is reduced, the variability in the expected value (for example, as measured by variance or standard deviation) is reduced. The ratio of this change can be captured and described by C_v. Consequently, a comparison of the relative merit of models in terms of their robustness can be represented by their respective values of C_v, in the sense that a model with a lower C_v is favored since there is less uncertainty associated with it. In fact, Markowitz (1952) advocates that the use of C_v as a measure of risk would equally ensure that the outcome of a decision-making process still lies in the set of efficient portfolios for the case of operational investments.

In a data set of normally distributed demands, if the C_v of demand is given as a case problem parameter, the standard deviation is computed by multiplication of C_v by the deterministic demand. Hence, increasing values of C_v result in increasing fluctuations in the demand and this is again undesirable.

Computation of C_v is based on the objective function of the formulated model. Table 6.1 displays the expressions to compute C_v for the proposed stochastic model formulations. Note that C_v for the deterministic case of each stochastic model should be equal to zero, by virtue of its standard deviation assuming a value of zero since it is based on the expected value solution.

Table 6.1 Determination of the coefficient of variation C_v for the deterministic and stochastic models.

Approach	Model	Objective function	Coefficient of variation $C_v = \dfrac{\sigma}{\mu} = \dfrac{\sqrt{V}}{E}$
Deterministic		$c^T x$	$C_v = 0$
1	Risk Model I	$\max z_1 = E[z_0] - \theta_1 V(z_0)$	$C_v = \dfrac{\sqrt{V(z_0)}}{E[z_0]}$
		or $\max z_1 = -V(z_0)$	
2	Expectation Models I and II	I: $\max z_2 = E[z_0] - \theta_1 V(z_0) - E_s$	$C_v = \dfrac{\sqrt{V(z_0)}}{E[z_0] - E_s}$
		II: $\max z_2 = -V(z_0) - E_s$	
3	Risk Model II	$\max z_3 = E[z_0] - \theta_1 V(z_0) - E_s - \theta_2 V_s$	$C_v = \dfrac{\sqrt{V(z_0) + V_s}}{E[z_0] - E_s}$
4	Risk Model III (MAD)	$\max z_3 = E[z_0] - \theta_1 V(z_0)$ $- E_s - \theta_3 W(p_s)$	$C_v = \dfrac{\sqrt{V(z_0) + W(p_s)}}{E[z_0] - E_s}$

6.6
Illustrative Case Study

We demonstrate the implementation of the proposed stochastic model formulations on the refinery planning linear programming (LP) model explained in Chapter 2. The original single-objective LP model is first solved deterministically and is then reformulated with the addition of the stochastic dimension according to the four proposed formulations. The complete scenario representation of the prices, demands, and yields is provided in Table 6.2.

The deterministic objective function of the LP model is given by:

$$\text{maximize } z = -8.0x_1 + 18.5x_2 + 8.0x_3 + 12.5x_4 + 14.5x_5 + 6.0x_6 - 1.5x_{14} \tag{6.29}$$

in which the negative coefficients denote the purchasing and operating costs while the positive coefficients are the sales prices of products. If c is the row vector of the price (or cost) and x is the column vector of production flow rate, then the objective function can be generally written as:

$$z = c^T x = \sum_{s \in S} \left(\sum_{i \in I} c_{i,s} x_i \right), \quad i = \{1,2,3,4,5,6,14\} \in I_{\text{price}}^{\text{random}} \subseteq I, \; s = \{1,2,3\} \in S \tag{6.30}$$

Table 6.2 Scenario formulation for uncertainty in prices, market demands, and product yields.

Product Type (i)	Scenario 1 Above average	Scenario 2 Average (expected value/mean)	Scenario 3 Below average		
Price uncertainty					
	Objective function coefficient of price, $c_{i,s}$ ($/ton)			Variance of price $v(c_{i,s})$ (($/ton)^2)	
Crude oil (1)	−8.8	−8.0	−7.2	0.352	
Gasoline (2)	20.35	18.5	16.65	1.882 375	
Naphtha (3)	8.8	8.0	7.2	0.352	
Jet fuel (4)	13.75	12.5	11.25	0.859 375	
Heating oil (5)	15.95	14.5	13.05	1.156 375	
Fuel oil (6)	6.6	6.0	5.4	0.198	
Cracker feed (14)	−1.65	−1.5	−1.35	0.012 375	
Demand uncertainty					
	Right-hand-side coefficient of constraints for production requirement (t/day)			Penalty cost incurred per unit ($/ton)	
				Shortfall in production ($c_{i,s}^+$)	Surplus in production ($c_{i,s}^-$)
Gasoline (2)	2835	2700	2565	25	20
Naphtha (3)	1155	1100	1045	17	13
Jet fuel (4)	2415	2300	2185	5	4
Heating oil (5)	1785	1700	1615	6	5
Fuel oil (6)	9975	9500	9025	10	8
Yield uncertainty					
	Left-hand-side coefficient of mass balances for fixed yields (unitless)			Penalty cost incurred per unit ($/unit)	
				Yield decrement ($q_{i,k,s}^+$)	Yield increment ($q_{i,k,s}^-$)
Naphtha (7)	−0.1365	−0.13	−0.1235	5	3
Jet fuel (4)	−0.1575	−0.15	−0.1425	5	4
Gas oil (8)	−0.231	−0.22	−0.209	5	3
Cracker feed (9)	−0.21	−0.20	−0.19	5	3
Residuum (10)	−0.265	−0.30	−0.335	5	3
Probability p_s	0.35	0.45	0.20		

6.6.1
Approach 1: Risk Model I

In this section we demonstrate the capability of our formulation in dealing with variations in the objective function prices, based on historical data. We present the uncertainty in terms of three scenarios: (i) the "above average" or optimistic scenario denoting a representative 10% positive deviation from the mean value; (ii) the "average" or realistic scenario that takes the expected values or mean; and (iii) the "below average" or pessimistic scenario, denoting a representative 10% negative deviation from the mean value.

The formulation for Risk Model I is:

$$\max z_1 = -V(z_0) = \sum_{s \in S} \left[\sum_{i \in I} p_s (c_{i,s} - \bar{c}_{i,s})^2 x_i^2 \right] \quad \text{(RM1)}$$

$$\text{subject to } E[z_0] = \sum_{s \in S} \left(\sum_{i \in I} p_s c_{i,s} x_i \right) \geq \text{Target profit value}$$

$$i = \{1, 2, 3, 4, 5, 6, 14\} \in I_{\text{price}}^{\text{random}} \subseteq I, \ s = \{1, 2, 3\} \in S$$

deterministic constraints (first stage) in the LP model.

As the main focus of this chapter is on the risk-incorporated models of Risk Models II and III, the computational results for Risk Model I are not presented here.

6.6.2
Approach 2: Expectation Models I and II

For simplicity of demonstration, it is assumed that there is no alternative source of production; hence, in a case of a shortfall in production, the demand is actually lost. Thus, the corresponding model considers the case where the in-house production of the refinery has to be anticipated at the beginning of the planning horizon.

A 5% standard deviation from the mean value of market demand for the saleable products in the LP model is assumed to be reasonable based on statistical analyses of the available historical data. To be consistent, the three scenarios assumed for price uncertainty with their corresponding probabilities are similarly applied to describe uncertainty in the product demands, as shown in Table 6.2, alongside the corresponding penalty costs incurred due to the unit production shortfalls or surpluses for these products. To ensure that the original information structure associated with the decision process sequence is respected, three new constraints to model the scenarios generated are added to the stochastic model. Altogether, this adds up to $3 \times 5 = 15$ new constraints in place of the five constraints in the deterministic model.

On the other hand, three representative scenarios are considered for modeling yield uncertainty for the left-hand side coefficient of fixed yields from the primary distillation unit. Each scenario corresponds to the depiction of "average product yield," "above average product yield," and "below average product yield," with a 5% deviation. To ensure satisfaction of the material balances, yields for the bottom product of crude unit are determined by subtracting the summation of yields for the other four products from unity. This does not distort the physics of the problem as the yield of residuum is relatively negligible anyway in a typical atmospheric distillation unit. In Chapter 8, we demonstrate a more general approach for handling yield uncertainty.

The penalty costs incurred per unit of shortages or excesses of crude oil yields are also shown in Table 6.2. The expectation Model I is formulated as:

6 Planning Under Uncertainty for a Single Refinery Plant

$$\max z_2 = \sum_{i \in I} \sum_{s \in S} p_s C_{i,s} x_i - \theta_1 \sum_{s \in S} \left[\sum_{i \in I} p_s (C_{i,s} - \bar{C}_{i,s})^2 x_i^2 \right]$$

$$- \sum_{i \in I} \sum_{s \in S} p_s \left[\begin{array}{c} (c_i^+ z_{i,s}^+ + c_i^- z_{i,s}^-) \\ + (q_i^+ y_{i,s}^+ + q_i^- y_{i,s}^-) \end{array} \right] \quad \text{(EM1')}$$

subject to: deterministic constraints (first stage) in the LP model,
stochastic constraints (second stage):

$$x_i + z_{i,s}^+ - z_{i,s}^- = d_{i,s}, \quad \forall i \in I, \forall s \in S \quad (6.31)$$

$$T_i x_1 + x_i + y_{i,k,s}^+ - y_{i,k,s}^- = 0, \quad i \in I, k \in K, s \in S \quad (6.32)$$

$$i = \{1, 2, 3, 4, 5, 6, 14\} \in I_{\text{price}}^{\text{random}} \subseteq I, s = \{1, 2, 3\} \in S.$$

The alternative Expectation Model II is expressed as:

$$\max z_2 = - \sum_{s \in S} \left[\sum_{i \in I} p_s (C_{i,s} - \bar{C}_{i,s})^2 x_i^2 \right] - \sum_{i \in I} \sum_{s \in S} p_s [(c_i^+ z_{i,s}^+ + c_i^- z_{i,s}^-) + (q_i^+ y_{i,s}^+ + q_i^- y_{i,s}^-)] \quad \text{(EM2')}$$

subject to: $E[z_0] = \sum_{s \in S} \left(\sum_{i \in I} p_s C_{i,s} x_i \right) \geq$ target profit value,

deterministic constraints in the LP model,
stochastic constraints (6.31) and (6.32),

$$i = \{1, 2, 3, 4, 5, 6, 14\} \in I_{\text{price}}^{\text{random}} \subseteq I, s = \{1, 2, 3\} \in S.$$

As in the case of Risk Model I, the computational results for Expectation Models I and II are not presented here as the emphasis of this chapter is on explaining the concept of risk analysis.

6.6.3
Approach 3: Risk Model II

The formulation of Risk Model II for the numerical example is given by the following:

$$\max z_3 = \sum_{i \in I} \sum_{s \in S} p_s C_i x_i - \theta_1 \sum_{i \in I} x_i^2 V(C_i) - \sum_{i \in I} \sum_{s \in S} p_s \left[\begin{array}{c} (c_i^+ z_{i,s}^+ + c_i^- z_{i,s}^-) \\ + (q_i^+ y_{i,s}^+ + q_i^- y_{i,s}^-) \end{array} \right]$$

$$- \theta_2 \sum_{s \in S} p_s \left\{ \begin{array}{c} \sum_{i \in I} [(c_i^+ z_{i,s}^+ + c_i^- z_{i,s}^-) + (q_i^+ y_{i,k,s}^+ + q_i^- y_{i,k,s}^-)] \\ - \sum_{i \in I} \sum_{s' \in S'} p_{s'} [(c_i^+ z_{i,s'}^+ + c_i^- z_{i,s'}^-) + (q_i^+ y_{i,k,s'}^+ + q_i^- y_{i,k,s'}^-)] \end{array} \right\}^2 \quad \text{(EM2')}$$

subject to : deterministic constraints in the LP model,
stochastic constraints (6.31) and (6.32),

$$i = \{1,2,3,4,5,6,14\} \in I_{price}^{random} \subseteq I, s = \{1,2,3\} \in S.$$

Tables 6.3–6.5 show the computational results for Risk Model II over a range of values of risk parameter θ_2 with respect to different recourse penalty costs, for three representative cases of $\theta_1 = 1E - 10$, $1E - 7$, and $1.55E - 5$, respectively. An example of the detailed results is presented in Table 6.6 for $\theta_2 = 50$ of the first case. Figure 6.2 illustrates the corresponding efficient frontier plot for Risk Model II while Figure 6.3 provides the plot of the expected profit for different levels of risk.

A number of different parameters are of interest in observing the robustness trends in both model and computed solution. Figure 6.3 shows that smaller values of θ_1 correspond to higher expected profit. With increasingly larger θ_1, the declining expected profit becomes almost constant and plateaus at \$81 770. The converse is also true with increasingly smaller θ_1 resulting in rising expected profit that eventually becomes roughly constant at the value of \$79 730.

Although increasing θ_2 with fixed value of θ_1 corresponds to decreasing expected profit, it generally leads to a reduction in expected production shortfalls and surpluses. Therefore, a suitable operating range of θ_2 values should be selected to achieve a proper trade-off between expected profit and expected production feasibility. Increasing θ_2 also reduces the expected deviation in the recourse penalty costs under different scenarios. This, in turn, translates to increased solution robustness. In that sense, the selection of θ_1 and θ_2 values depends primarily on the policy adopted by the decision maker.

In general, the coefficients of variation decrease with smaller values of θ_2. This is definitely desirable since it indicates that for higher expected profits there is diminishing uncertainty in the model, thus signifying model and solution robustness. It is also observed that for values of θ_2 approximately greater than or equal to 2, the coefficient of variation remain at a static value of 0.5237, thus indicating overall stability and a minimal degree of uncertainty in the model.

Since variance is a symmetric risk measure, profits both below and above the target levels are penalized equally, when it is actually desirable to only penalize the former instance. In other words, constraining or minimizing the variance of key performance metrics to achieve robustness may result in models that overcompensate for uncertainty, as highlighted by Samsatli, Papageorgiou, and Shah (1998). Eppen, Martin, and Schrage (1989) pointed out that another problem with using variance is that points on the MV efficient frontier may be stochastically dominated by other feasible solutions. A solution x_I is stochastically dominated by another solution x_{II} if for every scenario, the profit generated by x_{II} is at least as large as the profit given by x_I and yields a strictly greater profit for at least one scenario; a condition known as Pareto optimality in the multiobjective optimization literature Barbaro and Bagajewicz (2004). Ogryczak and Ruszczynski (2002) further explained that the mean-risk approach may produce inferior conclusions when typical dispersion statistics, such as variance, are employed as the risk measure.

Table 6.3 Representative computational results for Risk Model II for $\theta_1 = 1E-10$.

Operational risk factor θ_2	Optimal objective value	Expected variation in profit $V(z_0)(E+8)$	Expected total unmet demand/ production shortfall	Expected total excess production/ production surplus	Expected recourse penalty costs E_s	Expected variation in recourse penalty costs V_s	$\sigma = \sqrt{V(z_0) + V_s}$	Expected profit $E[z_0]$	$\mu = E[z_0] - E_s$	$C_v = \frac{\sigma}{\mu}$ Stochastic	Deterministic
1E−6	27 710	1.787	2575	26 820	54 060	8.167E+6	13 670	81 770	27 720	0.4932	(infeasible)
1E−4	26 900	1.787	2575	26 820	54 060	8.167E+6	13 670	81 770	27 720	0.4932	(infeasible)
1E−3	25 670	1.782	2467	26 710	53 840	2.047E+5	13 360	79 740	25 900	0.5157	(infeasible)
1E−2	25 510	1.782	2539	26 780	54 200	2047	13 350	79 730	25 530	0.5228	(infeasible)
1E−1	25 490	1.782	2518	26 760	54 230	20.47	13 350	79 730	25 490	0.5236	(infeasible)
1	25 490	1.782	2385	26 600	54 230	0.205	13 350	79 730	25 490	0.5236	(infeasible)
5	25 490	1.782	2508	26 750	54 230	0.008	13 350	79 730	25 490	0.5237	(infeasible)
10	25 490	1.782	2566	26 810	54 230	0.002	13 350	79 730	25 490	0.5237	(infeasible)
50	25 490	1.782	2556	26 800	54 230	8.189E−5	13 350	79 730	25 490	0.5237	(infeasible)
100	25 490	1.782	2519	26 800	54 230	2.047E−5	13 350	79 730	25 490	0.5237	(infeasible)
1000	25 490	1.782	2519	26 800	54 240	0	13 350	79 730	25 490	0.5237	(infeasible)
5000	25 490	1.782	2385	26 630	54 240	0	13 350	79 730	25 490	0.5237	(infeasible)
10 000	25 490	1.782	2542	26 780	54 240	0	13 350	79 730	25 490	0.5237	(infeasible)
10 500	25 490	1.782	2409	26 650	54 240	0	13 350	79 730	25 490	0.5237	(infeasible)
10 716	25 490	1.782	2508	26 750	54 240	0	13 350	79 730	25 490	0.5237	(infeasible)
10 717	(GAMS output: Infeasible solution. There are no superbasic variables.)										

Note: Trial solutions for $\theta_1 < 0.000\,001$ are not considered since improvement in expected profit is not anticipated based on trends in computed values.

6.6 Illustrative Case Study

Table 6.4 Representative computational results for Risk Model II for $\theta_1 = 1E-7$.

Operational risk factor θ_2	Optimal objective value	Expected variation in profit $V(z_0)(E+8)$	Expected total unmet demand/ production shortfall	Expected total excess production/ production surplus	Expected recourse penalty costs E_s	Expected variation in recourse penalty costs V_s	$\sigma = \sqrt{V(z_0) + V_s}$	Expected profit $E[z_0]$	$\mu = E[z_0] - E_s$	$C_v = \dfrac{\sigma}{\mu}$	
										Stochastic	Deterministic
1E−6	27 690	1.787	2575	26 820	54 060	8.167E+6	13 670	81 770	27 720	0.4932	(infeasible)
1E−4	26 880	1.787	2575	26 820	54 060	8.167E+6	13 670	81 770	27 720	0.4932	(infeasible)
1E−3	25 680	1.787	2467	26 710	53 840	2.047E+5	13 360	79 740	25 900	0.5157	(infeasible)
1E−2	25 490	1.787	2565	26 800	54 200	2047	13 350	79 730	25 530	0.5228	(infeasible)
1E−1	25 470	1.787	2385	26 620	54 230	20.47	13 350	79 730	25 490	0.5236	(infeasible)
1	25 470	1.787	2456	26 700	54 230	0.205	13 350	79 730	25 490	0.5236	(infeasible)
5	25 470	1.787	2519	26 760	54 230	0.008	13 350	79 730	25 490	0.5237	(infeasible)
10	25 470	17.82	2508	26 750	54 230	0.002	13 350	79 730	25 490	0.5237	(infeasible)
50	25 470	17.82	2385	26 630	54 230	8.189E−5	13 350	79 730	25 490	0.5237	(infeasible)
1E2	25 470	17.82	2385	26 630	54 230	2.047E−5	13 350	79 730	25 490	0.5237	(infeasible)
1E3	25 470	17.82	2508	26 750	54 240	0	13 350	79 730	25 490	0.5237	(infeasible)
5E3	25 470	17.82	2456	26 700	54 240	0	13 350	79 730	25 490	0.5237	(infeasible)
1E4	25 470	17.82	2475	26 720	54 240	0	13 350	79 730	25 490	0.5237	(infeasible)
1.05E4	25 470	17.82	2409	26 650	54 240	0	13 350	79 730	25 490	0.5237	(infeasible)
10 716	25 470	17.82	2584	26 820	54 240	0	13 350	79 730	25 490	0.5237	(infeasible)
10 717	(GAMS output: Infeasible solution. There are no superbasic variables.)										

Note: Trial solutions for $\theta_1 < 0.000001$ are not considered since improvement in expected profit is not anticipated based on trends in computed values.

Table 6.5 Representative computational results for Risk Model II for $\theta_1 = 1.55E-5$.

Operational risk factor θ_2	Optimal objective value	Expected variation in profit $V(z_0)(E+8)$	Expected total unmet demand/production shortfall	Expected total excess production/production surplus	Expected recourse penalty costs E_s	Expected variation in recourse penalty costs V_s	$\sigma = \sqrt{V(z_0)+V_s}$	Expected profit $E[z_0]$	$\mu = E[z_0] - E_s$	$C_v = \dfrac{\sigma}{\mu}$ Stochastic	Deterministic
1E−6	24 940	1.787	2575	26 820	54 060	8.167E+6	13 670	81 770	27 720	0.4932	(infeasible)
1E−4	24 130	1.787	2575	26 820	54 060	8.167E+6	13 670	81 770	27 720	0.4932	(infeasible)
2E−4	23 600	1.784	2435	26 670	53 590	3.010E+6	13 470	80 570	26 970	0.4993	(infeasible)
5E−4	23 140	1.782	2363	26 600	53 450	8.187E+5	13 380	79 760	26 310	0.5085	(infeasible)
1E−3	22 930	1.782	2493	26 730	53 840	2.047E+5	13 360	79 740	25 900	0.5157	(infeasible)
1E−2	22 750	1.782	2524	26 760	54 200	2047	13 350	79 730	25 530	0.5228	(infeasible)
1E−1	22 730	1.782	2531	26 770	54 230	20.470	13 350	79 730	25 490	0.5236	(infeasible)
1	22 730	1.782	2579	26 820	54 230	0.205	13 350	79 730	25 490	0.5236	(infeasible)
10	22 730	1.782	2580	26 820	54 230	0.002	13 350	79 730	25 490	0.5237	(infeasible)
50	22 730	1.782	2409	26 650	54 230	8.187E−5	13 350	79 730	25 490	0.5237	(infeasible)
1E2	22 730	1.782	2591	26 830	54 230	2.047E−5	13 350	79 730	25 490	0.5237	(infeasible)
5E2	22 730	1.782	2551	26 790	54 240	0	13 350	79 730	25 490	0.5237	(infeasible)
1E3	22 730	1.782	2532	26 770	54 240	0	13 350	79 730	25 490	0.5237	(infeasible)
5E3	22 730	1.782	2409	26 650	54 240	0	13 350	79 730	25 490	0.5237	(infeasible)
1E4	22 730	1.782	2561	26 800	54 240	0	13 350	79 730	25 490	0.5237	(infeasible)
1.2E4	22 730	1.782	2597	26 840	54 240	0	13 350	79 730	25 490	0.5237	(infeasible)
1.25E4	(GAMS output: Infeasible solution. A free variable exceeds the allowable range.)										

Note: Trial solutions for $\theta_1 < 0.000\,001$ are not considered since improvement in expected profit is not anticipated based on trends in computed values.

Table 6.6 Detailed computational results for Risk Model II for $\theta_1 = 1E-10$, $\theta_2 = 50$.

First-stage variable	Stochastic solution	Product (i)	Production shortfall z_{ij}^+ or surplus z_{ij}^- (t/d)					
			Scenario 1		Scenario 2		Scenario 3	
			z_{i1}^+	z_{i1}^-	z_{i2}^+	z_{i2}^-	z_{i3}^+	z_{i3}^-
x_1	15 000	Demands RHS coefficients randomness						
x_2	2000	Gasoline (2)	835.0	0	700.0	0	565.0	0
x_3	1155	Naphtha (3)	0	0	0	55.00	0	110.0
x_4	3638	Jet fuel (4)	0	1223	0	1338	0	1453
x_5	3598	Heating oil (5)	0	1813	0	1898	0	1983
x_6	9738	Fuel oil (6)	237.5	0	0	237.5	0	712.5
x_7	2155							
x_8	4635	Production yields LHS coefficients randomness						
x_9	4350	Naphtha (7)	0	107.5	0	205.0	0	302.5
x_{10}	5475	Jet fuel (4)	0	1275	0	1388	0	1500
x_{11}	1000	Gas oil (8)	0	1170	0	1335	0	1500
x_{12}	2698	Cracker feed (9)	0	1200	0	1350	0	1500
x_{13}	1937	Residuum (10)	0	1500	0	975.0	0	450.0
x_{14}	2500							
x_{15}	1850	E(Penalty Costs)	18 980	24 410	10 850			
x_{16}	1000	E_{total}	54 230					
x_{17}	1375							
x_{18}	899.4							
x_{19}	475.6							
x_{20}	125.0							
Expected profit z ($/day)	79 730							

To handle some of these shortcomings, Barbaro and Bagajewicz (2004) proposed a multiobjective optimization approach of simultaneous profit maximization and risk minimization for each profit target, which has been extensively applied in subsequent work by Bagajewicz and his coworkers (Aseeri and Bagajewicz, 2004; Bonfill et al., 2004). On the other hand, Eppen, Martin, and Schrage (1989) accounted for the expected downside risk in solving a real-world problem in the automobile industry. In his approach, a decision maker sets a target value for the desired profit; hence, the risk associated with a decision is measured by the failure to meet the target profit. A more general approach to ensure robustness was proposed by Samsatli, Papageorgiou, and Shah (1998), which can be tailored to various types of constraints to be imposed on the system and on specific suitable performance metrics. Other potentially more representative risk measures should also be considered with Kristoffersen (2005) providing a recent review of a wide choice of risk measures applicable within a two-stage stochastic programming framework.

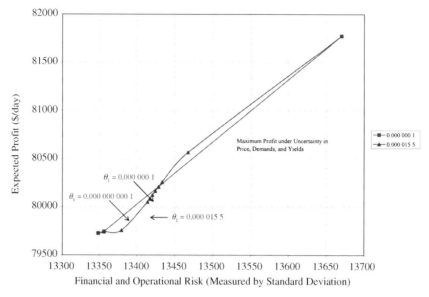

Figure 6.2 Risk Model II efficient frontier plot.

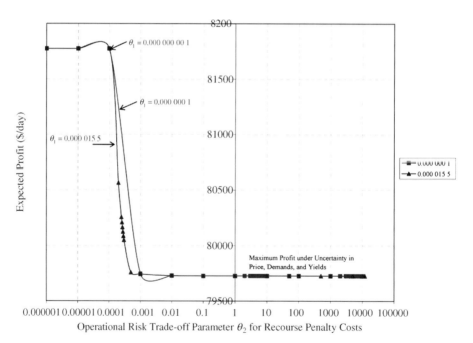

Figure 6.3 Risk Model II plot of expected profit for different levels of risk as represented by the economic risk factor θ_1 and the operational risk factor θ_2.

In the following section we illustrate how the mean-absolute deviation (MAD) is employed as the measure of operational risk.

6.6.4
Approach 4: Risk Model III

Risk Model III for the numerical example is formulated as follows:

$$\max z_4 = E(z_0) - \theta_1 V(z_0) - \sum_{i \in I}\sum_{s \in S} p_s[(c_i^+ z_{is}^+ + c_i^- z_{is}^-) + (q_i^+ y_{is}^+ + q_i^- y_{is}^-)]$$

$$-\theta_3 \sum_{s \in S} p_s \left| \sum_{i \in I} \begin{bmatrix} (c_i^+ z_{i,s}^+ + c_i^- z_{i,s}^-) \\ +(q_i^+ y_{i,s}^+ + q_i^- y_{i,s}^-) \end{bmatrix} - \sum_{i \in I}\sum_{s' \in S'} p_{s'} \begin{bmatrix} (c_i^+ z_{i,s'}^+ + c_i^- z_{i,s'}^-) \\ +(q_i^+ y_{i,k,s'}^+ + q_i^- y_{i,k,s'}^-) \end{bmatrix} \right| \quad \text{(RM3')}$$

subject to : deterministic constraints in the LP model,
stochastic constraints (6.31) and (6.32),
MAD linearization conditions (6.25) – (6.27).

From Table 6.7 and the corresponding efficient frontier plot in Figure 6.4, similar trends to Risk Model II (and also the expected value models) are observed in which decreasing values of θ_1 correspond to higher expected profit until a certain constant profit value is attained ($81 770). The converse is also true in which a constant profit of $59 330 is reached in the initially declining expected profit for increasing values of θ_1.

One of the reasons why the pair of decreasing values of θ_1 with a fixed value of θ_3 leads to increasing profit is due to the decrease in production shortfalls and, at the same time, increase in production surpluses. Typically, the fixed penalty cost for shortfalls is lower than surpluses. A good start would be to select a lower operating value of θ_1 to achieve both high model feasibility as well as increased profit. Moreover, lower values of θ_1 correspond to decreasing variation in the recourse penalty costs, which implies solution robustness.

6.7
General Remarks

In a general sense, sensitivity analysis is used to study the robustness of solutions to a linear programming (LP) model. However, in many cases, the output of sensitivity analysis could be misleading when used to assess the impact of uncertainty. As compared to stochastic programming, it would merely to help address the impact of uncertainty without providing the measures to hedge against it. As argued by many researchers, sensitivity analysis is a post-optimality analysis rather than a tool to account for uncertainty. On the other hand, stochastic programming is a constructive approach that is superior to sensitivity analysis. With stochastic linear programming (SLP) models, the decision maker is afforded the flexibility of introducing recourse variables to take corrective actions rather than reactive actions.

6 Planning Under Uncertainty for a Single Refinery Plant

Table 6.7 Representative computational results for Risk Model III for a selected representative range of values of θ_1 and θ_3.

Risk factors θ_1, θ_3	Optimal objective value	Expected deviation between profit $V(z_0)(E+7)$	Expected total unmet demand/ production shortfall	Expected total excess production/ production surplus	Expected recourse penalty costs E_s	Deviation in recourse penalty costs $W(p_s)$	$\sigma = \sqrt{V(z_0) + W(p_s)}$	Expected profit $E[z_0]$	$\mu = \|E[z_0] - E_s\|$	$C_v = \dfrac{\sigma}{\mu}$ Stochastic	Deterministic
0, 0	27 720	13.35	2575	26 820	54 060	2618	11 560	81 770	27 720	0.4169	(infeasible)
1E−6, 1E−4	27 580	13.35	2575	26 820	54 060	2618	11 560	81 770	27 720	0.4169	(infeasible)
1E−5, 1E−3	26 380	13.35	2575	26 820	54 060	2618	11 560	81 770	27 720	0.4169	(infeasible)
1E−4, 1E−3	14 360	13.35	2575	26 820	54 060	2618	11 560	81 770	27 720	0.4169	(infeasible)
1E−4, 1E−2	14 340	13.35	2575	26 820	54 060	2618	11 560	81 770	27 720	0.4169	(infeasible)
2E−4, 1E−2	2802	9.634	2575	17 800	40 230	2628	9815	62 330	22 100	0.4442	(infeasible)
3E−4, 1E−2	−6832	9.634	2575	17 800	40 230	2628	9815	62 330	22 100	0.4442	(infeasible)
5E−4, 1E−2	−25 310	8.837	2610	16 370	42 200	4253	9400	61 110	18 920	0.4970	(infeasible)
6E−4, 1E−2	−34 150	8.837	2610	16 370	42 200	4253	9400	61 110	18 920	0.4970	(infeasible)
7E−4, 1E−2	−42 800	7.966	2556	15 470	47 550	4256	8926	60 560	13 000	0.6863	(infeasible)
8E−4, 1E−2	−49 890	6.392	2424	13 670	58 200	4262	7995	59 450	1252	6.387	(infeasible)
9E−4, 1E−2	−56 170	6.237	2836	13 470	59 330	4263	7898	59 330	0	$\to \infty$	(infeasible)
1E−3, 1E−2	−62 410	6.237	2836	13 470	59 330	4263	7898	59 330	0	$\to \infty$	(infeasible)
1E−2, 0.1	−62 500	6.237	2836	13 470	59 330	4263	7898	59 330	0	$\to \infty$	(infeasible)
2E−2, 0.1	−1248 000	6.237	2836	13 470	59 330	4263	7898	59 330	0	$\to \infty$	(infeasible)
1, 1	−6.237E7	6.237	2836	13 470	59 330	4263	7898	59 330	0	$\to \infty$	(infeasible)
40, 20	−2.495E9	6.237	2836	13 470	59 330	4263	7898	59 330	0	$\to \infty$	(infeasible)
1E2, 10	−6.237E9	6.237	2836	13 470	59 330	4263	7898	59 330	0	$\to \infty$	(infeasible)
1E3, 1E2	−6.237E10	6.237	2836	13 470	59 330	4263	7898	59 330	0	$\to \infty$	(infeasible)
1E4, 1E2	(GAMS output: Infeasible solution. A free variable exceeds the allowable range.)										(infeasible)

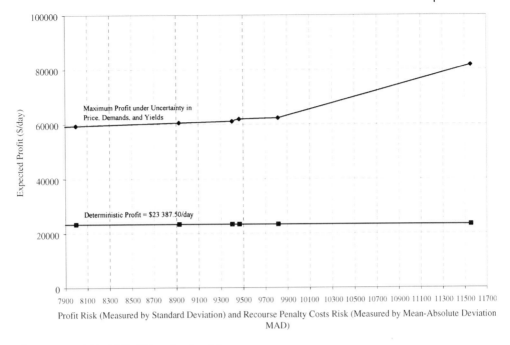

Figure 6.4 Risk Model III efficient frontier plot.

Nomenclature and Notation

Sets and Indices

I'	set of materials or products i
J	set of processes j
T	set of time periods t
S	set of scenarios s
K	set of products with yield uncertainty k

Deterministic Parameters

$d_{i,t,s}, d^L_{i,t,s}, d^U_{i,t,s}$	demand for product i in period t per realization of scenario s, with its corresponding constant lower (superscript L) and upper (superscript U) bounds
P_t	amount of crude oil purchased in period t
p^L_t, p^U_t	lower and upper bounds of the availability of crude oil during period t
$I^{f\ min}_{i,t}, I^{f\ max}_{i,t}$	minimum and maximum required amount of inventory for material i at the end of period t
$b_{i,j}$	stoichiometric coefficient for material i in process j
$\gamma_{i,t}$	unit sales price of product type i in period t
$\lambda_{i,t}$	unit purchase price of crude oil type i in period t

$\tilde{\lambda}_{i,t}$ value of the starting inventory of material i in period t
$\tilde{\gamma}_{i,t}$ value of the final inventory of material i in period t
$C_{j,t}$ operating cost of process j in period t
$h_{i,t}$ unit cost of subcontracting or outsourcing the production of product type i in period t
r_t, o_t cost per man-hour of regular and overtime labor in period t, respectively
$\alpha_{j,t}$ variable-size cost coefficient for the investment cost of capacity expansion of process j in period t
$\beta_{j,t}$ fixed-cost charge for the investment cost of capacity expansion of process j in period t
$\theta_1, \theta_2, \theta_3$ risk factors or weighting factors (weights) for multiobjective optimization procedure

Stochastic Parameters

p_s probability of scenario s
$\gamma_{i,s,t}$ unit sales price of product type i in period t per realization of scenario s
λ_t unit purchase price of crude oil in period t per realization of scenario s
$d_{i,s,t}$ demand for product i in time period t per realization of scenario s

Recourse Parameters

c_i^+ fixed penalty cost per unit demand $d_{i,s}$ of underproduction (shortfall) of product i per realization of scenario s (also the cost of lost demand)
c_i^- fixed penalty cost per unit demand $d_{i,s}$ of overproduction (surplus) of product i per realization of scenario s (also the cost of inventory to store production surplus)
$q_{i,k}^+$ fixed unit penalty cost for shortage in yields from material i for product k
$q_{i,k}^-$ fixed unit penalty cost for excess in yields from material i for product k

Deterministic Variables (First-Stage Decision Variables)

$x_{j,t}$ production capacity of process j during period t
$x_{j,t-1}$ production capacity of process j during period $t-1$
$CE_{j,t}, CE_{j,t}^L, CE_{j,t}^U$ capacity expansion of the plant for process j that is installed in period t, with its corresponding constant lower (superscript L) and upper (superscript U) bounds
$y_{j,t}$ binary decision variable that equals one (1) if there is an expansion for process j at the beginning of period t, and zero (0) otherwise
$S_{i,t}$ amount of product i sold in period t
$L_{i,t}$ amount of lost demand for product i in period t
P_t amount of crude oil purchased in period t
$I_{i,t}^s, I_{i,t}^f$ initial and final amount of inventory of material i in period t

$H_{i,t}$ amount of product type i to be subcontracted or outsourced in period t

R_t, O_t regular and overtime working or production hours in period t, respectively

Stochastic Recourse Variables (Second-Stage Decision Variables)

$z_{i,s}^+$ amount of unsatisfied demand for product i due to underproduction per realization of scenario s

$z_{i,s}^-$ amount of excess product i due to overproduction per realization of scenario s

$y_{i,k,s}^+$ amount of shortage in yields from material i for product type k per realization of scenario s

$y_{i,k,s}^-$ amount of excess in yields from material i for product type k per realization of scenario s

References

Acevedo, J. and Pistikopoulos, E.N. (1998) Stochastic optimization based algorithms for process synthesis under uncertainty. *Computers & Chemical Engineering*, **22**, 647.

Allen, D.H. (1971) Linear programming models for plant operations planning. *British Chemical Engineering*, **16**, 685.

Applequist, G.E., Samikoglu, O., Pekny, J., and Reklaitis, G.V. (1997) Issues in the use, design, and evolution of process scheduling and planning systems. *ISA Transactions*, **36**, 81.

Aseeri, A. and Bagajewicz, M.J. (2004) New measures and procedures to manage operational risk with applications to the planning of gas commercialization in Asia. *Computers & Chemical Engineering*, **28**, 2791.

Barbaro, A. and Bagajewicz, M.J. (2004) Managing operational risk in planning under uncertainty. *AIChE Journal*, **50**, 963.

Beale, E. (1955) On minimizing a convex function subject to linear inequalities. *Journal of the Royal Statistical Society, Series B (Methodological)*, **17**, 173.

Bernardo, F.P., Pistikopoulos, E.N., and Saraiva, P.M. (1999) Integration and computational issues in stochastic design and planning optimization problems. *Industrial & Engineering Chemistry Research*, **38**, 3056.

Bok, J.K., Lee, H., and Park, S. (1998) Robust investment model for long-range capacity expansion of chemical processing networks under uncertain demand forecast scenarios. *Computers & Chemical Engineering*, **22**, 1037.

Bonfill, A., Bagajewicz, M., Espuna, A., and Puigjaner, L. (2004) Risk management in the scheduling of batch plants under uncertain market demand. *Industrial & Engineering Chemistry Research*, **43**, 741.

Cheng, L., Subrahmanian, E., and Westerberg, A.W. (2005) Multiobjective decision processes under uncertainty: applications, problem formulations, and solution strategies. *Industrial & Engineering Chemistry Research*, **44**, 2405.

Dantzig, G.B. (1955) Linear programming under uncertainty. *Management Science*, **1**, 197.

Eppen, G.D., Martin, R.K., and Schrage, L. (1989) A scenario approach to capacity planning. *Operations Research*, **37**, 517.

Gupta, A. and Maranas, C.D. (2003) Managing demand uncertainty in supply chain planning. *Computers & Chemical Engineering*, **27**, 1219.

Ierapetritou, M.G. and Pistikopoulos, E.N. (1994) Novel optimization approach of stochastic planning models. *Industrial & Engineering Chemistry Research*, **33**, 1930.

Konno, H. and Koshizuka, T. (2005) Mean-absolute deviation model. *Institute of Industrial Engineers Transactions*, **37**, 893.

Konno, H. and Wijayanayake, A. (2002) Portfolio optimization under D.C. transaction costs and minimal transaction unit constraints. *Journal of Global Optimization*, **22**, 137.

Konno, H. and Yamazaki, H. (1991) Mean-absolute deviation portfolio optimization model and its applications to Tokyo Stock Market. *Management Science*, **37**, 519.

Kristoffersen, T.K. (2005) Deviation measures in linear two-stage stochastic programming. *Mathematical Methods of Operations Research*, **62**, 255.

Malcolm, S.A. and Zenios, S.A. (1994) Robust optimization of power systems capacity expansion under uncertainty. *Journal of Operational Research Society*, **45**, 1040.

Markowitz, H.M. (1952) Portfolio selection. *Journal of Finance*, **7**, 77.

Méndez, C.A., Cerda, J., Grossmann, I.E., Harjunkoski, I., and Fahl, M. (2006) State-of-the-art review of optimization methods for short-term scheduling of batch processes. *Computers & Chemical Engineering*, **30**, 913.

Mulvey, J.M., Vanderbei, R.J., and Zenios, S.A. (1995) Robust optimization of large-scale systems. *Operations Research*, **43**, 264.

Ogryczak, W. and Ruszczynski, A. (2002) Dual stochastic dominance and related mean-risk models. *SIAM Journal of Optimization*, **13**, 60.

Papahristodoulou, C. and Dotzauer, E. (2004) Optimal portfolios using linear programming models. *Journal of the Operational Research Society*, **55**, 1169.

Ravi, V. and Reddy, P.J. (1998) Fuzzy linear fractional goal programming applied to refinery operations planning. *Fuzzy Sets and Systems*, **96**, 173.

Samsatli, N.J., Papageorgiou, L.G., and Shah, N. (1998) Robustness metrics for dynamic optimization models under parameter uncertainty. *AIChE Journal*, **44**, 1993.

Shah, N. (1998) Single- and multisite planning and scheduling: current status and future challenges. AIChE Symposium Series: *Proceedings of the Third International Conference of the Foundations of Computer-Aided Process Operations, Snowbird, Utah, USA, July 5-10*, American Institute of Chemical Engineering, **94**, p. 75.

Swamy, C. and Shmoys, D.B. (2006) Algorithms column: approximation algorithms for two-stage stochastic optimization problems. *ACM SIGACT News*, **37**, 1.

Vin, J.P. and Ierapetritou, M.G. (2001) Robust short-term scheduling of multiproduct batch plants under demand uncertainty. *Industrial & Engineering Chemistry Research*, **40**, 4543.

7
Robust Planning of Multisite Refinery Network

In Chapter 3 of this book we discussed the problem of multisite refinery integration under deterministic conditions. In this chapter, we extend the analysis to account for different parameter uncertainty. Robustness is quantified based on both model robustness and solution robustness, where each measure is assigned a scaling factor to analyze the sensitivity of the refinery plan and integration network due to variations. We make use of the sample average approximation (SAA) method with statistical bounding techniques to generate different scenarios.

7.1
Introduction

Today the petroleum refining industry is facing a challenging task to remain competitive in a globalized market. High crude oil prices and growing stringent international protocols and regulations force petroleum companies to embrace every opportunity that increases their profit margin. A common solution is to seek integration alternatives, not only within a single facility but also on an enterprise-wide scale. This will provide enhanced utilization of resources and improved coordination and, therefore, achieve a global optimal production strategy within the network. However, considering such highly strategic planning decisions, particularly in the current volatile market leads to uncertainties playing a paramount role in the final decision making.

The remainder of this Chapter 7 is organized as follows. In Section 7.2 we will give a review of the related literature. Section 7.3 will present a model formulation for petroleum refining multisite network planning under uncertainty and using robust optimization. Then we will briefly explain the sample average approximation (SAA) method in Section 7.4. In Section 7.5, we will present computational results on industrial case studies consisting of a single refinery and a network of petroleum refineries. The chapter ends with concluding remarks in Section 7.6.

Planning and Integration of Refinery and Petrochemical Operations. Khalid Y. Al-Qahtani and Ali Elkamel
Copyright © 2010 WILEY-VCH Verlag GmbH & Co. KGaA, Weinheim
ISBN: 978-3-527-32694-5

7.2
Literature Review

Different approaches have been devised to tackle optimization under uncertainty including stochastic optimization (two-stage, multistage) with recourse based on the seminal work of Dantzig (1955), chance-constrained optimization (Charnes and Cooper, 1959), fuzzy programming (Bellman and Zadeh, 1970), and design flexibility (Grossmann and Sargent, 1978). These early works on optimization under uncertainty have undergone substantial developments in both theory and algorithms (Sahinidis, 2004). In this book, we employ stochastic programming with recourse which deals with problems with uncertain parameters of a given discrete or continuous probability distribution. The most common formulation of stochastic programming models for planning problems is the two-stage stochastic program. In a two-stage stochastic programming model, decision variables are cast into two groups: first stage and second stage variables. The first stage variables are decided upon prior to the actual realization of the random parameters. Once the uncertain events have unfolded, further design or operational adjustments can be made through values of the second-stage (alternatively called recourse variables at a particular cost). Stochastic programming with recourse commonly gives rise to large-scale models that require the use of decomposition methods and proper approximation techniques due to the high number of samples encountered (Liu and Sahinidis, 1996). However, recent developments in sampling techniques may help keep the stochastic program to a manageable size.

More recent applications and developments in the chemical engineering arena include the work by Ierapetritou and Pistikopoulos (1994) who proposed an algorithm for a two-stage stochastic linear planning model. The algorithm is based on design flexibility by finding a feasible subspace of the probability region instead of enumerating all possible uncertainty realizations. They also developed a Benders decomposition scheme for solving the problem without a priori discretization of the random space parameters. This was achieved by means of Gaussian quadrature numerical integration of the continuous density function. In a similar production planning problem, Clay and Grossmann (1997) developed a successive disaggregation algorithm for the solution of two-stage stochastic linear models with discrete uncertainty. Liu and Sahinidis 1995, 1996, 1997 studied the design uncertainty in process expansion using sensitivity analysis, stochastic programming and fuzzy programming, respectively. In their stochastic model, they used Monte Carlo sampling to calculate the expected objective function values. Their comparison of the different methodologies was in favor of stochastic models when the parameter distributions are not available. Ahmed, Sahinidis and Pistikopoulos (2000) proposed a modification to the decomposition algorithm of Ierapetritou and Pistikopoulos (1994). They were able to avoid solving the feasibility subproblems and instead of imposing constraints on the random space, they developed feasibility cuts on the master problem of their decomposition algorithm. The modification mitigates suboptimal solutions and develops a more accurate comparison to cost and flexibility. Neiro and Pinto (2005) developed a multiperiod MINLP model for production planning of refinery operations under uncertain petroleum and product prices and demand. They were able to solve the model for 19 periods and five scenarios.

Another stream of research considered risk and robust optimization. The representation of risk management using variance as a risk measure was proposed by Mulvey, Vanderbei and Zenios (1995) who referred to this approach as robust stochastic programming. They defined two types of robustness: (i) solution robustness referring to the optimal model solution when it remains close to optimal for any scenario realization, and (ii) model robustness representing an optimal solution when it is almost feasible for any scenario realization. Ahmed and Sahinidis (1998) proposed the use of an upper partial mean (UPM) as an alternative measure of variability with the aim of eliminating nonlinearities introduced by using variance. In addition to avoiding nonlinearity of the problem, UPM presents an asymmetric measure of risk, as opposed to variance, by penalizing unfavorable risk cases. Bok, Lee and Park (1998) proposed a multiperiod robust optimization model for chemical process networks with demand uncertainty and applied it to the petrochemical industry in South Korea. They adopted the robust optimization framework of Mulvey, Vanderbei and Zenios (1995) where they defined solution robustness as the model solution when it remains close to optimal for any demand realization, and model robustness when it has almost no excess capacity and unmet demand. More recently, Barbaro and Bagajewicz (2004) proposed a new risk metric to manage financial risk. They defined risk as the probability of not meeting a certain target profit, in the case of maximization, or cost, in the case of minimization. Additional binary variables are then defined for each scenario where each variable assumes a value of 1 in the case of not meeting the required target level; either profit or cost, and zero otherwise. Accordingly, appropriate penalty levels are assigned in the objective function. This approach mitigates the shortcomings of the symmetric penalization when using variance, but, on the other hand, adds computational burden through additional binary variables. Lin, Janak and Floudas (2004) proposed a robust optimization approach based on a min–max framework where they considered bounded uncertainty without known probability distribution. The uncertainty considered was in both the objective function coefficients and the right-hand-side of the inequality constraints and was then applied to a set of MILP problems. This approach allowed the violation of stochastic inequality constraints with a certain probability and uncertainty parameters were estimated from their nominal values through random perturbations. This approach, however, could result in large infeasibilities in some of the constraints when the nominal data values are slightly changed. This work was then extended by Janak, Lin and Floudas (2007) to cover known probability distributions and mitigate the large violations of constraints in Lin, Janak and Floudas (2004) via bounding the infeasibility of constraints and finding "better" nominal values of the uncertain parameters. It is worth mentioning that the work of both Lin, Janak and Floudas (2004) and Janak, Lin and Floudas (2007) is based on infeasibility/optimality analysis and does not consider recourse actions. For recent reviews on scheduling problems under uncertainty, we refer the interested reader to Janak, Lin and Floudas (2007), and for reviews on single and multisite planning and coordination to the review provided in see Chapter 3 of this book.

In this chapter, we extend the deterministic modeling for the design and analysis of multisite integration and coordination within a network of petroleum refineries proposed in Chapter 3 to consider uncertainty in raw materials and final product prices as well as products demand. The chapter considers both model robustness and solution robustness, following the Mulvey, Vanderbei and Zenios (1995) approach. The stochastic modeling consists of a two-stage stochastic MILP problem whereas the robust optimization is formulated as an MINLP problem with nonlinearity arising from modeling the risk components. We discuss parameter uncertainty in the coefficients of the objective function and the right-hand-side of inequality constraints. Furthermore, we describe how the sample average approximation (SAA) method within an iterative scheme is employed to generate required samples. Solution quality is statistically assessed by measuring the optimality gap of the final solution. The approach was applied to industrial scale case studies of a single petroleum refinery and a network of refineries.

7.3
Model Formulation

7.3.1
Stochastic Model

The formulation addresses the problem of determining an optimal integration strategy across multiple refineries and establishing an overall production and operating plan for each individual site. The deterministic model was explained in Chapter 3. In this chapter, uncertainty is accounted for using two-stage stochastic programming with recourse. Parameter uncertainties considered include uncertainties in the imported crude oil price $CrCost_{cr}$, product price Pr^{Ref}_{cfr}, (uncertainties in the coefficients of the objective function) and the market demand $D_{Ref\ cfr}$ (uncertainties in the right-hand-side of inequality constraints). Uncertainty is modeled through the use of mutually exclusive scenarios of the model parameters with a finite number N of outcomes. For each $\xi_k = (CrCost_{cr,k}, Pr^{Ref}_{cfr,k}, D_{Ref\ cfr,k})$ where $k = 1, 2, \ldots, N$, there corresponds a probability p_k. The generation of the scenarios as well as model statistical bounding will be explained in a later section. The stochastic model is given by:

$$\text{Min} \sum_{cr \in CR} \sum_{i \in I} \sum_{k \in N} p_k\, CrCost_{cr,k}\, S^{Ref}_{cr,i} + \sum_{p \in P} OpCost_p \sum_{cr \in CR} \sum_{i \in I} z_{cr,p,i}$$

$$+ \sum_{cir \in CIR} \sum_{i \in I} \sum_{i' \in I} InCost_{i,i'}\, y^{Ref}_{pipe\ cir,i,i'} + \sum_{i \in I} \sum_{m \in M_{Ref}} \sum_{s \in S} InCost_{m,s}\, y\,\exp^{Ref}_{m,i,s}$$

$$- \sum_{cfr \in PEX} \sum_{i \in I} \sum_{k \in N} p_k\, Pr^{Ref}_{cfr,k}\, e^{Ref}_{cfr,i} + \sum_{cfr \in CFR} \sum_{k \in N} p_k\, C^{Ref+}_{cfr}\, V^{Ref+}_{cfr,k}$$

$$+ \sum_{cfr \in CFR} \sum_{k \in N} p_k\, C^{Ref-}_{cfr}\, V^{Ref-}_{cfr,k} \quad \text{where} \quad i \neq i'$$

(7.1)

Subject to

$$z_{cr,p,i} = S^{Ref}_{cr,i} \quad \forall\, cr \in CR,\, i \in I \quad \text{where} \quad p \in P' = \{\text{Set of CDU processes } \forall\, \text{plant } i\} \tag{7.2}$$

$$\sum_{p \in P} \alpha_{cr,cir,i,p}\, z_{cr,p,i} + \sum_{i' \in I} \sum_{p \in P} \xi_{cr,cir,i',p,i}\, xi^{Ref}_{cr,cir,i',p,i}$$

$$- \sum_{i' \in I} \sum_{p \in P} \xi_{cr,cir,i,p,i'}\, xi^{Ref}_{cr,cir,i,p,i'} - \sum_{cfr \in CFR} w_{cr,cir,cfr,i} - \sum_{rf \in FUEL} w_{cr,cir,rf,i} = 0 \tag{7.3}$$

$$\forall\, cr \in CR,\, cir \in CIR,\, i' \,\&\, i \in I \quad \text{where} \quad i \neq i'$$

$$\sum_{cr \in CR} \sum_{cir \in CB} w_{cr,cir,cfr,i} - \sum_{cr \in CR} \sum_{rf \in FUEL} w_{cr,cfr,rf,i} = x^{Ref}_{cfr,i} \quad \forall\, cfr \in CFR,\, i \in I \tag{7.4}$$

$$\sum_{cr \in CR} \sum_{cir \in CB} \frac{w_{cr,cir,cfr,i}}{sg_{cr,cir}} = xv^{Ref}_{cfr,i} \quad \forall\, cfr \in CFR,\, i \in I \tag{7.5}$$

$$\sum_{cir \in FUEL} cv_{rf,cir,i}\, w_{cr,cir,rf,i} + \sum_{cfr \in FUEL} w_{cr,cfr,rf,i} - \sum_{p \in P} \beta_{cr,rf,i,p}\, z_{cr,p,i} = 0$$

$$\forall\, cr \in CR,\, rf \in FUEL,\, i \in I \tag{7.6}$$

$$\sum_{cr \in CR} \sum_{cir \in CB} \left(att_{cr,cir,q \in Qv} \frac{w_{cr,cir,cfr,i}}{sg_{cr,cir}} + att_{cr,cir,q \in Qw} \left[w_{cr,cir,cfr,i} - \sum_{rf \in FUEL} w_{cr,cfr,rf,i} \right] \right)$$

$$\geq q^{L}_{cfr,q \in Qv}\, xv^{Ref}_{cfr,i} + q^{L}_{cfr,q \in Qw}\, x^{Ref}_{cfr,i} \quad \forall\, cfr \in CFR,\, q = \{Qw, Qv\},\, i \in I \tag{7.7}$$

$$\sum_{cr \in CR} \sum_{cir \in CB} \left(att_{cr,cir,q \in Qv} \frac{w_{cr,cir,cfr,i}}{sg_{cr,cir}} + att_{cr,cir,q \in Qw} \left[w_{cr,cir,cfr,i} - \sum_{rf \in FUEL} w_{cr,cfr,rf,i} \right] \right)$$

$$\leq q^{U}_{cfr,q \in Qv}\, xv^{Ref}_{cfr,i} + q^{U}_{cfr,q \in Qw}\, x^{Ref}_{cfr,i} \quad \forall\, cfr \in CFR,\, q = \{Qw, Qv\},\, i \in I \tag{7.8}$$

$$\text{Min } C_{m,i} \leq \sum_{p \in P} \gamma_{m,p} \sum_{cr \in CR} z_{cr,p,i} \leq \text{Max } C_{m,i} + \sum_{s \in S} AddC_{m,i,s}\, y^{Ref}_{\exp m,i,s} \quad \forall\, m \in M_{Ref},\, i \in I \tag{7.9}$$

$$\sum_{cr \in CR} \sum_{p \in P} \xi_{cr,cir,i,p,i'}\, xi^{Ref}_{cr,cir,i,p,i'} \leq F^{U}_{cir,i,i'}\, Y^{Ref}_{pipe\, cir,i,i'} \quad \forall\, cir \in CIR,\, i' \,\&\, i \in I \quad \text{where} \quad i \neq i' \tag{7.10}$$

$$\sum_{i \in I} \left(x^{\text{Ref}}_{cfr,i} - e^{\text{Ref}}_{cfr',i} \right) + V^{\text{Ref}+}_{cfr,k} - V^{\text{Ref}-}_{cfr,k} = D_{\text{Ref } cfr,k} \quad \forall\, cfr \in CFR \quad cfr' \in PEX \quad k \in N \tag{7.11}$$

$$IM^L_{cr} \leq \sum_{i \in I} S^{\text{Ref}}_{cr,i} \leq IM^U_{cr} \quad \forall\, cr \in CR \tag{7.12}$$

The above formulation is a two-stage stochastic mixed-integer linear programming (MILP) model. Objective function (7.1) minimizes the first stage variables and the penalized second stage variables. Similar to the analysis of inventory problems (Ahmed, Çakman and Shapiro, 2007), the production over the target demand is penalized as an additional inventory cost of each ton of refined products. Similarly, shortfall in a certain product demand is assumed to be satisfied at the product spot market price. The recourse variables $V^{\text{Ref}+}_{cfr,k}$ and $V^{\text{Ref}-}_{cfr,k}$ in Equation 7.11 represent the shortfall and surplus for each random realization $k \in N$, respectively. These will compensate for the violations in Equation 7.11 and will be penalized in the objective function using appropriate shortfall and surplus costs $C^{\text{Ref}+}_{cfr}$ and $C^{\text{Ref}-}_{cfr}$, respectively. Uncertain parameters are assumed to follow a normal distribution for each outcome of the random realization ξ_k. Although this might sound restrictive, this assumption brings no limitation to the generality of the proposed approach as other distributions can easily be used instead. The recourse variables $V^{\text{Ref}+}_{cfr,k}$ and $V^{\text{Ref}-}_{cfr,k}$ in this formulation will compensate for deviations from the mean of the market demand.

7.3.2
Robust Model

The above stochastic model takes a decision merely based on first-stage and expected second-stage costs leading to an assumption that the decision-maker is risk-neutral. The generic representation of risk can be written as:

$$\text{Min}_{x,y}\, c^T x + E[Q(x, \xi(\omega))] + \lambda f(\omega, y)$$

where f is a measure of variability of the second-stage costs and λ is a non-negative scalar representing risk tolerance which is usually decided by the modeler. This representation is referred to as a mean-risk model (Ahmed, Çakman and Shapiro, 2007). This formulation follows the representation of the Markowitz mean-variance (MV) model (Markowitz, 1952). The variability measure can be modeled as variance, mean-absolute deviation or financial measures of value-at-risk (VaR) and conditional value-at-risk (CVaR).

Risk is modeled in terms of variance in both prices of imported crude oil $CrCost_{cr}$ and petroleum products Pr^{Ref}_{cfr}, represented by first stage variables, and forecasted demand $D_{\text{Ref } cfr}$, represented by the recourse variables. The variability in the prices represents the solution robustness in which the model solution will remain close to optimal for all scenarios. On the other hand, variability of the recourse term represents the model robustness in which the model solution will almost be feasible for all scenarios. This technique gives rise to a multiobjective optimization problem in which scaling factors are used to evaluate the sensitivity due to variations of each term. The variations in the raw material and product prices are scaled by θ_1 and the deviation from forecasted demand is scaled by θ_2. Different values of θ_1 and θ_2 are used in order to observe the sensitivity of each term on the final solution of the problem. The objective function with risk consideration can be written as shown in (7.13):

$$\begin{aligned}
\text{Min} \quad & \sum_{cr \in CR} \sum_{i \in I} \sum_{k \in N} p_k \, CrCost_{cr,k} \, S_{cr,i}^{Ref} + \sum_{p \in P} OpCost_p \sum_{cr \in CR} \sum_{i \in I} z_{cr,p,i} \\
& + \sum_{cir \in CIR} \sum_{i \in I} \sum_{i' \in I} InCost_{i,i'} \, y_{pipe\,cir,i,i'}^{Ref} + \sum_{i \in I} \sum_{m \in M_{Ref}} \sum_{s \in S} InCost_{m,s} \, y_{exp\,m,i,s}^{Ref} \\
& - \sum_{cfr \in PEX} \sum_{i \in I} \sum_{k \in N} p_k \, Pr_{cfr,k}^{Ref} \, e_{cfr,i}^{Ref} + \sum_{cfr \in CFR} \sum_{k \in N} p_k \, C_{cfr}^{Ref+} \, V_{cfr,k}^{Ref+} \\
& + \sum_{cfr \in CFR} \sum_{k \in N} p_k \, C_{cfr}^{Ref-} \, V_{cfr,k}^{Ref-} + \theta_1 \left[\sqrt{\text{var}\left(CrCost_{cr,k} \, S_{cr,i}^{Ref}\right)} + \sqrt{\text{var}\left(Pr_{cfr,k}^{Ref} \, e_{cfr,i}^{Ref}\right)} \right] \\
& + \theta_2 \left[\sqrt{\text{var}\left(C_{cfr}^{Ref+} \, V_{cfr,k}^{Ref+}\right)} + \sqrt{\text{var}\left(C_{cfr}^{Ref-} \, V_{cfr,k}^{Ref-}\right)} \right]
\end{aligned}$$

(7.13)

By expanding the mean and variance terms of $CrCost_{cr,k}$, $Pr_{cfr,k}^{Ref}$, $V_{cfr,k}^{Ref+}$ and $V_{cfr,k}^{Ref-}$, the objective function (7.13) can be recast as:

$$\begin{aligned}
\text{Min} \quad & \sum_{cr \in CR} \sum_{i \in I} \sum_{k \in N} p_k \, CrCost_{cr,k} \, S_{cr,i}^{Ref} + \sum_{p \in P} OpCost_p \sum_{cr \in CR} \sum_{i \in I} z_{cr,p,i} \\
& + \sum_{cir \in CIR} \sum_{i \in I} \sum_{i' \in I} InCost_{i,i'} \, y_{pipe\,cir,i,i'}^{Ref} + \sum_{i \in I} \sum_{m \in M_{Ref}} \sum_{s \in S} InCost_{m,s} \, y_{exp\,m,i,s}^{Ref} \\
& - \sum_{cfr \in PEX} \sum_{i \in I} \sum_{k \in N} p_k \, Pr_{cfr,k}^{Ref} \, e_{cfr,i}^{Ref} + \sum_{cfr \in CFR} \sum_{k \in N} p_k \, C_{cfr}^{Ref+} \, V_{cfr,k}^{Ref+} \\
& + \sum_{cfr \in CFR} \sum_{k \in N} p_k \, C_{cfr}^{Ref-} \, V_{cfr,k}^{Ref-} \\
& + \theta_1 \left(\sqrt{\sum_{cr \in CR} \sum_{i \in I} \sum_{k \in N} \left(S_{cr,i}^{Ref}\right)^2 p_k \left[CrCost_{cr,k} - \sum_{k \in N} p_k \, CrCost_{cr,k}\right]^2} \right. \\
& \left. + \sqrt{\sum_{cr \in CR} \sum_{i \in I} \sum_{k \in N} \left(e_{cfr,i}^{Ref}\right)^2 p_k \left[Pr_{cfr,k}^{Ref} - \sum_{k \in N} p_k \, Pr_{cfr,k}^{Ref}\right]^2} \right) \\
& + \theta_2 \left(\sqrt{\sum_{cfr \in CFR} \sum_{k \in N} \left[C_{cfr}^{Ref+}\right]^2 p_k \left[V_{cfr,k}^{Ref+} - \sum_{k \in N} p_k \, V_{cfr,k}^{Ref+}\right]^2} \right. \\
& \left. + \sqrt{\sum_{cfr \in CFR} \sum_{k \in N} \left[C_{cfr}^{Ref-}\right]^2 p_k \left[V_{cfr,k}^{Ref-} - \sum_{k \in N} p_k \, V_{cfr,k}^{Ref-}\right]^2} \right)
\end{aligned}$$

(7.14)

In order to understand the effect of each term on the overall objective function of the system, different values of θ_1 and θ_2 are evaluated to construct the efficient frontier of expected cost versus risk measured by standard deviation. This will be demonstrated in the illustrative case studies.

7.4
Sample Average Approximation (SAA)

7.4.1
SAA Method

The solution of stochastic problems is generally very challenging as it involves numerical integration over the random continuous probability space of the second stage variables (Goyal and Ierapetritou, 2007). An alternative approach is the discretization of the random space using a finite number of scenarios. This approach has received increasing attention in the literature since it gives rise to a deterministic equivalent formulation which can then be solved using available optimization algorithms. A common approach is Monte Carlo sampling where independent pseudo-random samples are generated and assigned equal probabilities (Ruszczyński and Shapiro, 2003).

The use of numerical integration through Gaussian quadratures and cubatures was studied by Pistikopoulos and Ierapetritou (1995). Acevedo and Pistikopoulos (1996) compared the Gaussian numerical integration methods with the Monte Carlo sampling technique and suggested the use of cubature methods for smaller dimensional problems and sampling-based methods for larger problems. The sampling-based methods were further classified by Verweij et al. (2003) to either "interior" or "exterior" sampling. In the interior sampling approach, samples can be adjusted during the optimization procedure either by adding additional samples to the previously generated ones, taking subsets of the samples, or even by generating completely new samples. Examples of this approach include the stochastic decomposition algorithm by Higle and Sen (1991) and the branch and bound algorithm by Norkin, Pflug and Ruszczysk (1998). On the other hand, exterior sampling includes the class of problems where samples are generated "outside" the optimization algorithm and then the samples are used to construct and solve the problem as a deterministic equivalent. The sample average pproximation (SAA) method, also known as stochastic counterpart, is an example of an exterior sampling approach. The sample average approximation problem can be written as (Verweij et al., 2003):

$$v_N = \min_{x \in X} c^T x + \frac{1}{N} \sum_{k \in N} Q(x, \xi^k) \tag{7.15}$$

It approximates the expectation of the stochastic formulation (usually called the "true" problem) and can be solved using deterministic algorithms. The SAA method was used among others by Shapiro and Homem-de-Mello (1998), Mark, Morton and

Wood (1999), Linderoth, Shapiro and Wright (2002) for stochastic linear problems, Kleywegt, Shapiro and Homem-De-Mello (2001), Verweij et al. (2003) for stochastic integer problems, and Wei and Realff (2004), Goyal and Ierapetritou (2007) for MINLP problems. Problem (7.15) can be solved iteratively in order to provide statistical bounds on the optimality gap of the objective function value. For details and proofs see Norkin, Pflug and Ruszczysk (1998) and Mark, Morton and Wood (1999). The procedure consists of a number of steps as described in the following section.

7.4.2
SAA Procedure

- Generate R independent sample batches (denoting the number of sample replication) each with sample size of N, that is, $\xi^{j1}, \ldots, \xi^{jN}, j = 1, \ldots, R$. For each sample size N solve the sample average approximation problem defined as:

$$v_N^j = \min_{x \in X} c^T x + \frac{1}{N} \sum_{k \in N} Q(x, \xi^{kj}) \quad (7.16)$$

The objective values $v_N^1, \ldots v_N^R$ of problem (7.16) and their corresponding solutions $\hat{x}_N^1, \ldots \hat{x}_N^R$ are then obtained.

- Calculate:

$$\bar{v}_N = \frac{1}{R} \sum_{j \in R} v_N^j \quad (7.17)$$

$$\frac{\sigma_{\bar{v}_N}^2}{R} = \frac{1}{R(R-1)} \sum_{j \in R} (v_N^j - \bar{v}_N)^2 \quad (7.18)$$

According to Mark, Morton and Wood (1999) and Norkin, Pflug and Ruszczysk (1998), the value of $\bar{v}_N$ in (7.17) is less than or equal to the true optimal value v^* obtained by solving the "true" problem, see Appendix C for proof. Therefore, $\bar{v}_N$ is a statistical lower bound to the true optimal value with a variance estimator of $\frac{\sigma_{\bar{v}_N}^2}{R}$ calculated by Equation 7.18.

- Select any candidate solution $\hat{x}_N^1, \ldots, \hat{x}_N^R$ obtained from the previous steps. However, it is preferable to select the optimal solution with the minimum objective function value, that is:

$$\hat{x}^* \in \arg \min [v_N^R : \hat{x}_N^1, \hat{x}_N^2, \ldots, \hat{x}_N^R] \quad (7.19)$$

Fix the solution value to the point obtained from the above minimization in (7.19); generate an independent sample $N' = \xi^1, \ldots, \xi^{N'}$ and compute the value of the following objective function:

$$\hat{v}_{N'} = \min_{\hat{x}^*} c^T \hat{x}^* + \frac{1}{N'} \sum_{k \in N'} Q(\hat{x}^*, \xi^k) \quad (7.20)$$

Considering the relatively less computational effort required to solve problem (7.20), the value of N' is typically chosen to be quite larger than N in order to obtain an accurate estimation of $\hat{v}_{N'}$ (Verweij et al., 2003). Since $\hat{x}^*$ is a feasible point to the true problem, we have $\hat{v}_{N'} \geq v^*$. Hence, $\hat{v}_{N'}$ is a statistical upper bound to the true problem with a variance estimated by Equation 7.21:

$$\frac{\sigma^2_{\hat{v}_{N'}}}{N'} = \frac{1}{N'(N'-1)} \sum_{k \in N'} (c^T \hat{x}^* + Q(\hat{x}^*, \xi^k) - \hat{v}_{N'})^2 \qquad (7.21)$$

From the above procedure, we can estimate the $(1-\alpha)$ confidence interval of the optimal gap. For a given $t_{n-1,\alpha/2}$ where t is the critical value of the t-distribution with $(n-1)$ degrees of freedom, the following can be estimated:

$$\tilde{\varepsilon}_l = t_{n-1,\alpha/2} \frac{\sigma_{\bar{v}_N}}{\sqrt{R}} \quad \text{and} \quad \tilde{\varepsilon}_u = t_{n-1,\alpha/2} \frac{\sigma_{\hat{v}_{N'}}}{\sqrt{N'}}$$

Then, the optimality gap can be constructed as:

$$[0, \{\hat{v}_{N'} - \bar{v}_N\}^+ + \tilde{\varepsilon}_u + \tilde{\varepsilon}_l] \quad \text{where} \quad \{y\}^+ \equiv \max(y, 0). \qquad (7.22)$$

Note that due to sampling error we may find that $\hat{v}_{N'} < \bar{v}_N$. For this reason the confidence interval obtained by (7.22) provides a more conservative bounding. The above procedure for the validation of a candidate solution was originally suggested by Norkin, Pflug and Ruszczysk (1998) and further developed by Mark, Morton and Wood (1999).

7.5
Illustrative Case Study

This section presents computational results of the models and the sampling scheme proposed in this chapter. The refinery examples considered represent industrial-scale size refineries and an actual configuration that can be found in many industrial sites around the world. In the presentation of the results, we focus on demonstrating the sample average approximation computational results as we vary the sample sizes and compare their solution accuracy and the CPU time required for solving the models.

The modeling system GAMS (Brooke et al., 1996) is used for setting up the optimization models. The computational tests were carried out on a Pentium M processor 2.13 GHz. The models were solved with DICOPT (Viswanathan and Grossmann, 1990). The NLP subproblems were solved with CONOPT2 (Drud, 1994), while the MILP master problems were solved with CPLEX (CPLEX Optimization Inc, 1993).

7.5.1
Single Refinery Planning

This example illustrates the performance of the proposed approach on a single site total refinery planning problem. The refinery scale, capacity and configuration mimic an existing refinery in the Middle East. Figure 7.1 is a state equipment network (SEN)

7.5 *Illustrative Case Study* | 149

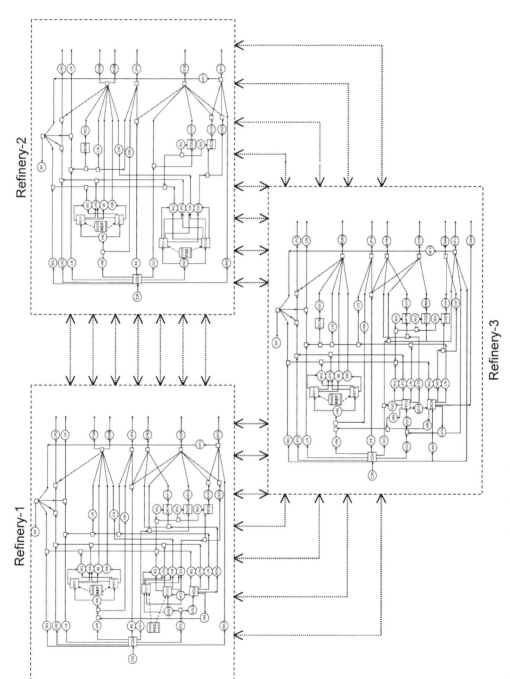

Figure 7.1 SEN representation of a multiple refineries integration network.

representation of a multiple refineries network, where in this example we will study Refinery 1. The refinery uses a single feedstock (Arabian Light) to feed the atmospheric crude unit where it separates crude oil into several fractions including LPG, naphtha, kerosene, gas oil and residues. The heavy residues are then sent to the vacuum unit where they are further separated into vacuum gas oil and vacuum residues. In general, depending on the production targets, different processing and treatment processes are applied to the crude fractions. In our case, the naphtha is further separated into heavy and light naphtha. Heavy naphtha is sent to the catalytic reformer unit to produce high octane reformates for gasoline blending and light naphtha is sent to the light naphtha pool and to an isomerization unit to produce isomerate for gasoline blending too.

The middle distillates are combined with other similar intermediate streams and sent for hydrotreating and then for blending to produce jet fuels and gas oils. Atmospheric and vacuum gas oils are further treated by either fluid catalytic cracking (FCC) or hydrocracking (HC) in order to increase the gasoline and distillate yields. These distillates from both FCC and HC are desulfurized in the cycle gas oil desulfurization and ATK desulfurization processes. The final products in this example consist of liquefied petroleum gas (LPG), light naphtha (LT), two grades of gasoline (PG98 and PG95), No.4 jet fuel (JP4), military jet fuel (ATKP), No.6 gas oil (GO6), and heating fuel oil (HFO).

The major capacity limitations as well as the availability constraints are shown in Table 7.1. Raw materials, product prices, and demand uncertainty were assumed to

Table 7.1 Major capacity constraints of single refinery planning.

	Higher limit (1000 t/year)
Production Capacity	
Distillation	12 000
Reforming	2000
Fluid catalytic cracker	1000
Hydrocracker	2000
Des gas oil	3000
Des cycle gas oil	100
Des ATK	1200
Crude availability	
Arabian Light	12 000
Local demand	
LPG	$\mathcal{N}$ (320,20)
LN	$\mathcal{N}$ (220,20)
PG98	$\mathcal{N}$ (50,5)
PG95	$\mathcal{N}$ (1600,20)
JP4	$\mathcal{N}$ (1300,20)
GO6	$\mathcal{N}$ (2500,50)
ATKP	$\mathcal{N}$ (500,20)
HFO	$\mathcal{N}$ (700,20)

follow a normal distribution. However, this assumption brings no restriction to the generality of the proposed modeling approach as other sampling distributions can easily be used instead. Prices of crude oil and refined products reflect the current market prices and assume a standard deviation of $10 US. The problem is formulated as an LP since there is no integration or capacity expansion requirement in this case study.

Table 7.2 and Figure 7.2 show different confidence interval values of the optimality gap when changing the sample sizes N and N', while fixing the number of replications R to 30. The replication number R need not be very large since usually 5–7 replications are sufficient to get an idea about the variability of $\bar{v}_N$ (Qian and Shapiro, 2006). It can be seen that increasing the sample size N has more weight on reducing the optimality gap, and, therefore, the variability of the objective function, than increasing N'.

However, the increase in the sample size N will depend on computational time and available computer memory. In our particular case studies, we run into memory limitations when we increase the sample size N beyond 2000 samples. Table 7.3 shows the solution of the single refinery problem using the SAA scheme with $N = 2000$ and $N' = 20000$. The proposed approach required 553 CPU s to converge to the optimal solution.

The single refinery was then solved considering risk in terms of variations in the price of imported crude oil, prices of final products and forecasted demand to provide a

Table 7.2 Computational results with SAA of single refinery planning.

U B sample size	Number of samples: $R = 30$	Lower bound sample size = N 500	1000	1500	2000
$N' = 5000$	LB estimate: $\bar{v}_N$	2 888 136	2 888 660	2 888 964	2 888 978
	LB error: $\tilde{\varepsilon}_l(\alpha = 0.975)$	1834	1446	1050	1036
	UB estimate: $\hat{v}_{N'}$	2 889 367	2 889 076	2 889 720	2 887 545
	UB error: $\tilde{\varepsilon}_u(\alpha = 0.975)$	3636	3693	3653	3637
	95% Conf. interval	[0,6701]	[0,5555]	[0,5458]	[0,4673]
	CPU (s)	26	33	41	49
$N' = 10\,000$	LB estimate: $\bar{v}_N$	2 887 536	2 889 941	2 888 123	2 888 526
	LB error: $\tilde{\varepsilon}_l(\alpha = 0.975)$	2367	1311	1045	1332
	UB estimate: $\hat{v}_{N'}$	2 887 864	2 890 973	2 888 884	2 888 706
	UB error: $\tilde{\varepsilon}_u(\alpha = 0.975)$	2574	2608	2601	2619
	95% Conf. interval	[0,5269]	[0,4951]	[0,4407]	[0,4131]
	CPU (s)	109	113	117	128
$N' = 20\,000$	LB estimate: $\bar{v}_N$	2 888 639	2 889 023	2 888 369	2 888 293
	LB error: $\tilde{\varepsilon}_l(\alpha = 0.975)$	2211	1604	1190	1090
	UB estimate: $\hat{v}_{N'}$	2 889 593	2 889 033	2 888 527	2 888 311
	UB error: $\tilde{\varepsilon}_u(\alpha = 0.975)$	1841	1847	1835	1840
	95% Conf. interval	[0,5006]	[0,3462]	[0,3183]	[0,2949]
	CPU (s)	534	536	545	553

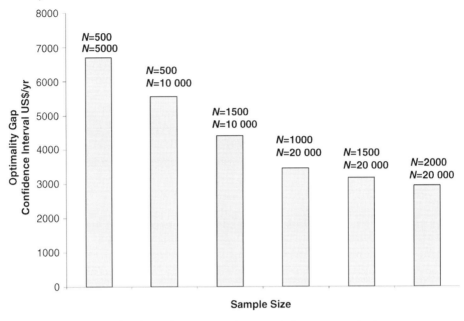

Figure 7.2 Single refinery planning optimality gap variations with sample size.

more robust analysis of the problem. The problem was formulated as an NLP problem with nonlinearity due to modeling risk in terms of variance. As mentioned earlier, the problem will have a more robust solution as the results will remain close to optimal for all scenarios through minimizing variations of the raw material and product prices. In

Table 7.3 Model results of single refinery planning.

Process variables		Results (1000 t/year)
Production levels	Crude unit	12 000
	Reformer 95	
	Reformer 100	1824
	FCC gasoline mode	96
	FCC gas oil mode	708
	Hydrocracker	1740
	Des gas oil	2891
	Des cycle gas oil	45
	Des ATK	1200
Exports	PG95	463
	JP4	502
	GO6	1293
	ATKP	1036
	HFO	1253
Total cost ($/yr)		2 887 687

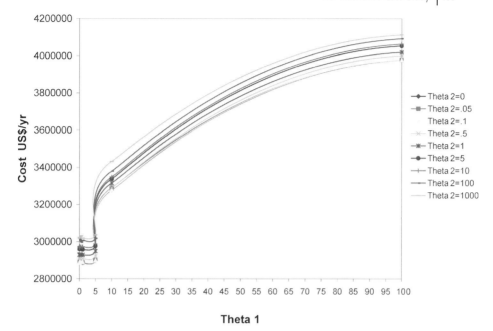

Figure 7.3 Cost variation with different values of θ_1 and θ_2 for single refinery planning.

a similar analogy, the model will be more robust as we are minimizing the variations of the recourse variables (demand) leading to a solution that is almost feasible for all scenarios. The effect of each robustness term was evaluated by varying their scaling factors. The model was repeatedly solved for different values of θ_1 (price variations) and θ_2 (demand variations) in order to construct the efficient frontier plot of expected cost versus risk measured by standard deviation.

Figure 7.3 illustrates the change in cost with respect to different values of θ_1 (price variations) and θ_2 (demand variations). The cost tends to increase as higher scaling values are given to the standard deviation of prices and demand. It can be seen from the graph that the problem shows more sensitivity to variations in raw material and product prices than to the variations in demand. Figure 7.4 illustrates the trade-off between the expected cost vs. risk, represented by the total standard deviation of prices and demand. The expected cost decreases at higher values of risk in terms of price and demand variations with respect to different θ_1 and θ_2 values. Generally, the values of θ_1 and θ_2 will depend on the policy adopted by the investor or the plant operator whether they are risk-averse or risk takers, and can be read directly from the efficient frontier plots.

7.5.2
Multisite Refinery Planning

In this example, we extend the scale of the case study to cover strategic planning for three complex refineries by which we demonstrate the performance of our model to

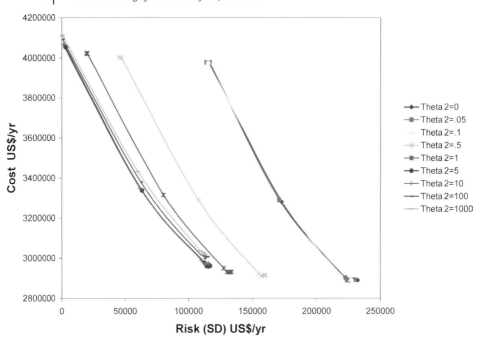

Figure 7.4 Trade-off between cost and risk at different θ_2 values while varying θ_1 for single refinery planning.

devise an overall production plan and an integration strategy. See Figure 7.1 for the overall topology of the refineries and Table 7.4 for major capacity constraints. The three refineries are assumed to be in one industrial area, which is a common situation in many locations around the world, and are coordinated through a main headquarters sharing the feedstock supply. The cost parameters for pipeline installation were calculated as cost per distance between the refineries, and then multiplied by the required pipe length in order to connect any two refineries. The pipeline diameter considered in all cases was 8 inches. The final products of the three refineries consist of liquefied petroleum gas (LPG), light naphtha (LT), two grades of gasoline (PG98 and PG95), No. 4 jet fuel (JP4), military jet fuel (ATKP), No.6 gas oil (GO6), diesel fuel (Diesel), heating fuel oil (HFO), and petroleum coke (coke). This problem was formulated as an MILP with the overall objective of minimizing total annualized cost.

Similar to the single refinery planning example in Section 7.5.1, the problem was solved for different sample sizes N and N' to illustrate the variation of optimality gap confidence intervals, as shown in Table 7.5 and Figure 7.5. The results illustrate the trade-off between model solution accuracy and computational effort. Furthermore, the increase in the sample size N has a more pronounced effect on reducing the optimality gap, however, due to computa-

Table 7.4 Major refineries capacity constraints for multisite refinery planning.

Production Capacity	Higher limit (1000 t/year)		
	R1	R2	R3
Distillation	4500	12 000	9900
Reforming	1000	2000	1800
Isomerization	200	—	450
Fluid catalytic cracker	800	1200	—
Hydrocracker	—	2000	2500
Delayed coker	—	—	1500
Des gas oil	1300	3000	2400
Des cycle gas oil	200	750	—
Des ATK	—	1200	1680
Des distillates	—	—	350
Crude availability			
Arabian Light	31 200		
Local demand			
LPG	N (432,20)		
LN	N (250,20)		
PG98	N (540,20)		
PG95	N (4440,50)		
JP4	N (2340,50)		
GO6	N (4920,50)		
ATK	N (1800,50)		
HFO	N (200,20)		
Diesel	N (400,20)		
Coke	N (300,20)		

tional and memory limitations we did not increase the sample size N beyond 2000 samples.

Table 7.6 shows the solution of the refineries network using the SAA scheme with $N = 2000$ and $N' = 20000$ where the proposed model required 790 CPU s to converge to the optimal solution. In addition to the master production plan devised for each refinery, the solution proposed the amounts of each intermediate stream to be exchanged between the different processes in the refineries. The formulation considered the uncertainty in the imported crude oil prices, petroleum product prices and demand. The three refineries collaborate to satisfy a given local market demand where the model provides the production and blending level targets for the individual sites. The annual production cost across the facilities was found to be $6 650 868.

When considering risk, in an analogous manner to the single refinery case, the problem was solved for different values of θ_1 and θ_2 to construct the efficient

7 Robust Planning of Multisite Refinery Network

Table 7.5 Computational results with SAA of the multisite refinery planning.

UB sample Size	Number of Samples: $R = 30$	Lower bound sample size = N			
		500	1000	1500	2000
$N' = 5000$	LB estimate: $\bar{v}_N$	6 647 411	6 647 713	6 647 569	6 647 109
	LB error: $\tilde{\varepsilon}_l(\alpha = 0.975)$	4752	3185	2667	2516
	UB estimate: $\hat{v}_{N'}$	6 661 697	6 658 832	6 656 480	6 652 869
	UB error: $\tilde{\varepsilon}_u(\alpha = 0.975)$	7877	7850	7746	7769
	95% Conf. interval	[0,26915]	[0,22154]	[0,19325]	[0,16044]
	CPU (s)	38	44	60	69
$N' = 10\,000$	LB estimate: $\bar{v}_N$	6 647 660	6 649 240	6 648 294	6 650 178
	LB error: $\tilde{\varepsilon}_l(\alpha = 0.975)$	3868	3750	2280	2105
	UB estimate: $\hat{v}_{N'}$	6 656 195	6 658 611	6 656 763	6 655 654
	UB error: $\tilde{\varepsilon}_u(\alpha = 0.975)$	5447	5550	5463	5550
	95% Conf. interval	[0,17850]	[0,18671]	[0,16213]	[0,13132]
	CPU (s)	147	158	160	173
$N' = 20\,000$	LB estimate: $\bar{v}_N$	6 647 296	6 649 123	6 649 634	6 649 421
	LB error: $\tilde{\varepsilon}_l(\alpha = 0.975)$	3924	3248	2173	1720
	UB estimate: $\hat{v}_{N'}$	6 656 006	6 656 778	6 656 684	6 654 684
	UB error: $\tilde{\varepsilon}_u(\alpha = 0.975)$	3923	3924	3895	3879
	95% Conf. interval	[0,16557]	[0,14828]	[0,13118]	[0,10862]
	CPU (s)	748	763	770	790

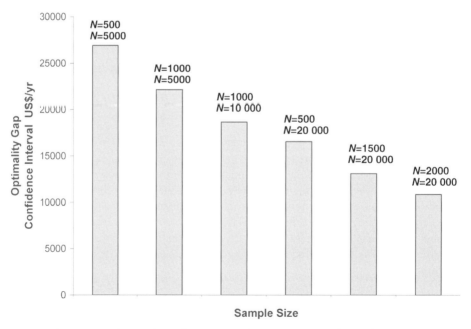

Figure 7.5 Multisite refinery planning optimality gap variations with sample size.

Table 7.6 Model results of the multisite refinery planning.

Process variables				Results (1000 t/year)		
				R1	R2	R3
Crude oil supply	Arabian Light			4500	12 000	9900
Production levels	Crude unit			4500	12 000	9900
	Reformer 95			172	539	—
	Reformer 100			401	1285	1772
	Isomerization			200	—	450
	FCC gasoline mode			616	1033	—
	FCC gas oil mode			—	—	—
	Hydrocracker			—	1740	2436
	Delayed coker			—	—	1100
	Des gas oil			1035	2760	2378
	Des cycle gas oil			200	536	—
	Des ATK			—	1200	1680
	Des distillates			—	—	264
Intermediate streams exchange	From	R1	VGO	—	—	338 to HCU
			CGO	—	75 to DCGO	—
		R2	LN	60 to Isom.	—	295 to Isom.
		R3	UCO	—	134 to FCC	—
Exports	PG95			296		
	JP4			1422		
	GO6			3657		
	HFO			1977		
	ATK			1837		
	Coke			40		
Total cost ($/yr)				6 650 868		

frontier plot. Figure 7.6 illustrates how cost increases with respect to higher values of θ_1 and θ_2. The cost of operating the multisite refinery network will depend on the scaling values assigned to prices and demand variations. Figure 7.7 demonstrates the trade-off between the cost and the total standard deviation of both prices and demand, denoted as risk. The figures show that the cost of production and designing the integration network between the refineries is more sensitive to variations in crude cost and product prices when compared to demand variations.

Furthermore, for values of θ_1 and θ_2 exceeding 100, the model did not recommend exchange of intermediate streams between the refineries due to the high risk associated with such investment. However, the values of both θ_1 and θ_2 are left to the decision maker's preference.

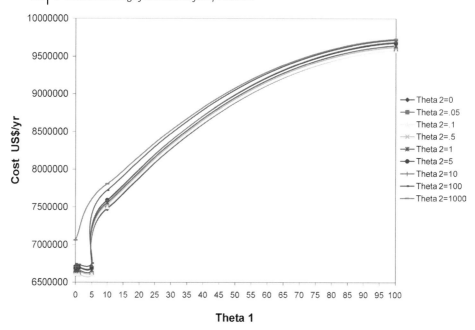

Figure 7.6 Cost variation with different values of θ_1 and θ_2 for multisite refinery planning.

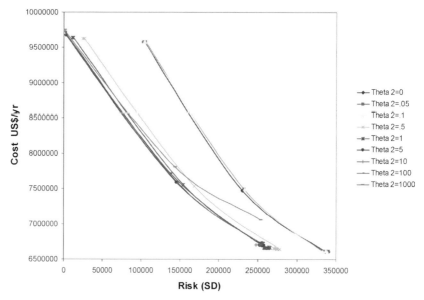

Figure 7.7 Trade-off between cost and risk at different θ_2 values while varying θ_1 for multisite refinery planning.

7.6
Conclusion

In this chapter, we proposed a two-stage stochastic MILP model to design an integration strategy under uncertainty and plan capacity expansions, as required, in a multisite refinery network. The proposed method employs the sample average approximation (SAA) method with a statistical bounding and validation technique. In this sampling scheme, a relatively small sample size N is used to make decisions, with multiple replications, and another independent larger sample is used to reassess the objective function value while fixing the first stage variables and solving for the second stage variables. In addition, robust optimization of the refinery network integration model was also evaluated. The proposed approach led to results that are more stable against variability in imported crude oil and product prices (solution robustness) as well as forecasted product demand (model robustness). Furthermore, the study showed that the refinery models are more sensitive to variations in the prices of imported crude oil and exported final products as opposed to variations in product demand. The scaling values of the solution and model robustness depend on the policy adopted by the investor, whether being risk-averse or a risk taker, and can be read directly from the efficient frontier plots presented in this chapter.

References

Acevedo, J. and Pistikopoulos, E.N. (1996) Computational studies of stochastic optimization algorithms for process synthesis under uncertainty. *Computers & Chemical Engineering*, **20**, S1.

Ahmed, S. and Sahinidis, N.V. (1998) Robust process planning under uncertainty. *Industrial & Engineering Chemistry Research*, **37**, 1883.

Ahmed, S., Çakman, U., and Shapiro, A. (2007) Coherent risk measure in inventory problems. *European Journal of Operational Research*, **182**, 226.

Ahmed, S., Sahinidis, N.V., and Pistikopoulos, E.N. (2000) An improved decomposition algorithm for optimization under uncertainty. *Computers & Chemical Engineering*, **23**, 1589.

Barbaro, A. and Bagajewicz, M.J. (2004) Managing financial risk in planning under uncertainty. *AIChE Journal*, **5**, 963.

Bellman, R. and Zadeh, L.A. (1970) Decision-making in a fuzzy environment. *Management Science*, **17**, 141.

Bok, J., Lee, H., and Park, S. (1998) Robust investment model for long-range capacity expansion of chemical processing networks under uncertain demand forecast scenarios. *Computers & Chemical Engineering*, **22**, 1037.

Brooke, A., Kendrick, D., Meeraus, A., and Raman, R. (1996) *GAMS–A User's Guide*, GAMS Development Corporation, Washington DC.

Charnes, A. and Cooper, W. (1959) Chance-constrained programming. *Management Science*, **6**, 73.

Clay, R.L. and Grossmann, I.E. (1997) A disaggregation algorithm for the optimization of stochastic planning models. *Computers & Chemical Engineering*, **21**, 751.

CPLEX Optimization, Inc . (1993) *Using the CPLEX Callable Library and CPLEX Mixed Integer Library*, CPLEX Optimization Inc., Incline Village, NV.

Dantzig, G.B. (1955) Linear programming under uncertainty. *Management Science*, **1**, 197.

Drud, A.S. (1994) CONOPT: a large scale GRG code. *ORSA Journal of Computing*, **6**, 207.

Goyal, V. and Ierapetritou, M.G. (2007) Stochastic MINLP optimization using simplicial approximation. *Computers and Chemical Engineering*, **31**, 1081.

Grossmann, I.E. and Sargent, R.W.H. (1978) Optimum designing of chemical plants with uncertain parameters. *AIChE Journal*, **24**, 1021.

Higle, J.L. and Sen, S. (1991) Stochastic decomposition: an algorithm for two stage linear programs with recourse. *Mathematics of Operations Research*, **16**, 650.

Ierapetritou, M.G. and Pistikopoulos, E.N. (1994) Novel optimization approach of stochastic planning models. *Industrial & Engineering Chemistry Research*, **33**, 1930.

Janak, S.L., Lin, X., and Floudas, C.A. (2007) A new robust optimization approach for scheduling under uncertainty: II. Uncertainty with known probability distribution. *Computers & Chemical Engineering*, **31**, 171.

Kleywegt, A.J., Shapiro, A., and Homem-De-Mello, T. (2001) The sample average approximation method for stochastic discrete optimization. *SIAM Journal on Optimization*, **12**, 479.

Lin, X., Janak, S.L., and Floudas, C.A. (2004) A new robust optimization approach for scheduling under uncertainty: I. Bounded uncertainty. *Computers & Chemical Engineering*, **28**, 1069.

Linderoth, J., Shapiro, A., and Wright, S. (2002) The empirical behavior of sampling methods for stochastic programming. Optimization online, http://www.optimization-online.org/DB_HTML/2002/01/424.html.

Liu, M.L. and Sahinidis, N.V. (1995) Computational trends and effects of approximations in an MILP model for process planning. *Industrial & Engineering Chemistry Research*, **34**, 1662.

Liu, M.L. and Sahinidis, N.V. (1996) Optimization in process planning under uncertainty. *Industrial & Engineering Chemistry Research*, **35**, 4154.

Liu, M.L. and Sahinidis, N.V. (1997) Process planning in a fuzzy environment. *European Journal of Operational Research*, **100**, 142.

Mark, W.K., Morton, D.P., and Wood, R.K. (1999) Monte Carlo bounding techniques for determining solution quality in stochastic programs. *Operational Research Letters*, **24**, 47.

Markowitz, H.M. (1952) Portfolio selection. *Journal of Finance*, **7**, 77.

Mulvey, J.M., Vanderbei, R.J., and Zenios, S.A. (1995) Robust optimization of large-scale systems. *Operations Research*, **43**, 264.

Neiro, S.M.S. and Pinto, J.M. (2005) Multiperiod optimization for production planning of petroleum refineries. *Chemical Engineering Communications*, **192**, 62.

Norkin, W.I., Pflug, G.Ch., and Ruszczysk, A. (1998) A branch and bound method for stochastic global optimization. *Mathematical Programming*, **83**, 425.

Pistikopoulos, E.N. and Ierapetritou, M.G. (1995) Novel approach for optimal process design under uncertainty. *Computers & Chemical Engineering*, **19**, 1089.

Qian, Z. and Shapiro, A. (2006) Simulation-based approach to estimation of latent variable models. *Computational Statistics & Data Analysis*, **51**, 1243.

Ruszczyński, R. and Shapiro, A. (2003) *Handbooks in Operations Research and Management Science-Volume 10: Stochastic Programming*, Elsevier Science, Amsterdam, Netherlands.

Sahinidis, N.V. (2004) Optimization under un-certainty: state-of-the-art and opportunities. *Computers & Chemical Engineering*, **28**, 971.

Shapiro, A. and Homem-de-Mello, T. (1998) A simulation-based approach to two stage stochastic programming with recourse. *Mathematical Programming*, **81**, 301.

Verweij, B., Ahmed, S., Kleywegt, A.J., Nemhauser, G., and Shapiro, A. (2003) The sample average approximation method applied to stochastic routing problems: A computational study. *Computational Optimization & Applications*, **24**, 289.

Viswanathan, J. and Grossmann, I.E. (1990) A combined penalty function and outer-approximation method for MINLP optimization. *Computers & Chemical Engineering*, **14**, 769.

Wei, J. and Realff, M.J. (2004) Sample average approximation methods for stochastic MINLPs. *Computers & Chemical Engineering*, **28**, 336.

8
Robust Planning for Petrochemical Networks

This chapter addresses the planning, design and optimization of a network of petrochemical processes under uncertainty and robust considerations. Similar to the previous chapter, robustness is analyzed based on both model robustness and solution robustness. Parameter uncertainty includes process yield, raw material and product prices, and lower product market demand. The expected value of perfect information (EVPI) and the value of the stochastic solution (VSS) are also investigated to illustrate numerically the value of including the randomness of the different model parameters.

8.1
Introduction

The discussions on planning under uncertainty in Chapter 7 and planning in the petrochemical networks earlier in Chapter 4 underline the importance of modeling uncertainty and considering risk in process system engineering studies. In this chapter, we extend the model presented in Chapter 4 to address the strategic planning, design and optimization of a network of petrochemical processes under uncertainty and robust considerations. Robustness is analyzed based on both model robustness and solution robustness, where each measure is assigned a scaling factor to analyze the sensitivity of the integration network to variations. The stochastic model is formulated as a two-stage stochastic MILP problem whereas the robust optimization is formulated as an MINLP problem with nonlinearity arising from modeling the risk components. Both endogenous uncertainty, represented by uncertainty in the process yield, and exogenous uncertainty, represented by uncertainty in raw material and product prices, and lower product market demand are considered. The concept of expected value of perfect information (EVPI) and the value of the stochastic solution (VSS) are also investigated to illustrate numerically the value of including randomness of the different model parameters. The consideration of uncertainty in these parameters provided a more robust and practical analysis of the problem, especially at a time when fluctuations in petroleum and petrochemical products prices and demands are soaring. For a literature review on

Planning and Integration of Refinery and Petrochemical Operations. Khalid Y. Al-Qahtani and Ali Elkamel
Copyright © 2010 WILEY-VCH Verlag GmbH & Co. KGaA, Weinheim
ISBN: 978-3-527-32694-5

planning of petrochemical networks, planning under uncertainty, and modeling risk, we refer the reader to Chapters 4 and 7.

The remainder of this Chapter is organized as follows. In Section 8.2 we will discuss model formulation for petrochemical network planning under uncertainty and with uncertainty and risk consideration, referred to as robust optimization. Then we will briefly explain the concept of value of information and stochastic solution, in Section 8.3. In Section 8.4, we will illustrate the performance of the model through an industrial case study. The chapter ends with concluding remarks in Section 8.5.

8.2
Model Formulation

8.2.1
Two-Stage Stochastic Model

This formulation is an extension to the deterministic model presented earlier in this book. In Chapter 4 all parameters of the model were assumed to be known with certainty. However, the current situation of fluctuating high petroleum crude oil and petrochemical product prices and demands is an indication of the high market and industry volatility. Acknowledging the shortcomings of deterministic models, parameter uncertainty is considered in the process yield $\delta_{cp,m}$, raw material and product prices Pr_{cp}^{Pet}, and lower product demand $D_{Pet\,cp}^{L}$. The problem is formulated as a two-stage stochastic programming model. The uncertainty is considered through discrete distribution of the random parameters with a finite number S of possible outcomes (scenarios) $\xi_k = (Pr_{cp,k}^{Pet}, \delta_{cp,m,k}, D_{Pet\,cp,k}^{L})$ corresponding to a probability p_k. The formulation of the stochastic model is as follows:

$$\text{Max} \quad \sum_{cp \in CP} \sum_{m \in M_{\text{Pet}}} \sum_{k \in N} p_k \, Pr_{cp,k}^{Pet} \, \delta_{cp,m,k} \, x_m^{Pet}$$

$$- \sum_{cp \in CFP} \sum_{k \in N} p_k \, C_{cp}^{Pet+} \, V_{cp,k}^{Pet+} - \sum_{cp \in CFP} \sum_{k \in N} p_k \, C_{cp}^{Pet-} \, V_{cp,k}^{Pet-} \quad (8.1)$$

Subject to

$$F_{cp}^{Pet} + \sum_{m \in M_{\text{Pet}}} \delta_{cp,m} \, x_m^{Pet} + V_{cp \in CFP,k}^{Pet+} - V_{cp,k}^{Pet-} = D_{Pet\,cp,k}^{L} \quad \forall cp \in CP \quad k \in N \quad (8.2)$$

$$F_{cp}^{Pet} + \sum_{m \in M_{\text{Pet}}} \delta_{cp,m} \, x_m^{Pet} \leq D_{Pet\,cp \in CFP}^{U} \quad \forall cp \in CP \quad k \in N \quad (8.3)$$

$$B_m^L \, y_{proc\,m}^{Pet} \leq x_m^{Pet} \leq K^U \, y_{proc\,m}^{Pet} \quad \forall m \in M_{\text{Pet}} \quad (8.4)$$

$$\sum_{cp \in CIP} Y^{Pet}_{proc\ m} \leq 1 \quad \forall\ m \in M_{Pet} \text{ that produces } cp \in CIP \text{ (intermediate)}$$
(8.5)

$$\sum_{cp \in CFP} Y^{Pet}_{proc\ m} \leq 1 \quad \forall\ m \in M_{Pet} \text{ that produces } cp \in CFP \text{ (final)} \quad (8.6)$$

$$F^{Pet}_{cp} \leq S^{Pet}_{cp} \quad \forall\ cp \in CP \quad (8.7)$$

The above formulation is a two-stage mixed-integer linear programming (MILP) model. The recourse variables $V^{Pet+}_{cp,k}$ and $V^{Pet-}_{cp,k}$ represent the shortfall and surplus for each random realization $k \in S$, respectively. These will compensate for the violations in constraints (8.2) and will be penalized in the objective function using the appropriate shortfall and surplus costs C^{Pet+}_{cp} and C^{Pet-}_{cp}, respectively. Uncertain parameters are assumed to follow a normal distribution for each outcome of the random realization ξ_k. The scenarios for all random parameters are generated simultaneously. The recourse variables $V^{Pet+}_{cp,k}$ and $V^{Pet-}_{cp,k}$ in this formulation will compensate for deviations from the mean of the lower market demands $D^L_{Pet\ cp}$ and process yield $\delta_{cp,m}$. In this way, the use of an independent recourse action and compensation for the violation in the constraint due to process yield uncertainty alone is avoided. Although this may not allow for an explicit analysis for process yield uncertainty, it instead circumvents the complication of treating these types of endogenous uncertainties.

8.2.2
Robust Optimization

The stochastic model with recourse in the previous section takes a decision merely based on first-stage and expected second-stage costs leading to an assumption that the decision-maker is risk-neutral (Sahinidis, 2004). In order to capture the concept of risk in stochastic programming, Mulvey, Vanderbei and Zenios (1995) proposed the following amendment to the objective function:

$$\text{Min}_{x,y}\ c^T x + E[Q(x,\xi(\omega))] + \lambda f(\omega, y)$$

where $E[Q(x,\xi(\omega))]$ is the fixed recourse, f is a measure of variability (i.e., second moment) of the second-stage costs, and λ is a non-negative scalar representing risk tolerance. The representation through risk management using variance as a risk measure is often referred to as robust stochastic programming (Mulvey, Vanderbei and Zenios, 1995). This is also a typical risk measure following the Markowitz mean-variance (MV) model (Markowitz, 1952). The robustness is incorporated through the consideration of higher moments (variance) of the random parameter distribution ξ_k in the objective function, and, hence, measuring the trade-offs between mean value and variability.

In this study, operational risk was accounted for in terms of variance in both projected benefits, represented by first stage variables, and forecasted demand, represented by the recourse variables. The variability in the projected benefit represents the solution robustness where the model solution will remain close to optimal for all scenarios. On the other hand, variability of the recourse term represents the model robustness where the model solution will almost be feasible for all scenarios. This approach gives rise to a multiobjective analysis in which scaling factors are used to evaluate the sensitivity due to variations in each term. The projected benefits variation was scaled by θ_1, and deviation from forecasted demand was scaled by θ_2, where different values of θ_1 and θ_2 were used in order to observe the sensitivity of each term on the final petrochemical complex. The objective function with risk consideration can be written as follows:

$$\text{Max} \sum_{cp \in CP} \sum_{m \in M_{Pet}} \sum_{k \in N} p_k \, Pr^{Pet}_{cp,k} \, \delta_{cp,m,k} \, x^{Pet}_m$$

$$- \sum_{cp \in CFP} \sum_{k \in N} p_k \, C^{Pet+}_{cp} \, V^{Pet+}_{cp,k} - \sum_{cp \in CFP} \sum_{k \in N} p_k \, C^{Pet-}_{cp} \, V^{Pet-}_{cp,k} \quad (8.8)$$

$$-\theta_1 \sqrt{\text{var}(\text{Profit uncertainty})} - \theta_2 \sqrt{\text{var}(\text{Recourse uncertainty})}$$

Since the randomness in the profit uncertainty term is a product of two random parameters, process yield $\delta_{cp,m,k}$ and chemical prices $Pr^{Pet}_{cp,k}$, its variance can be written based on the variance of a product of two variables x and y (Johnson and Tetley, 1955), that is:

$$\text{var}_{xy} = \text{var}_x \, \text{var}_y + \text{var}_x \, \mu_y + \text{var}_y \, \mu_x$$

where var_x and μ_x represent the variance and mean value of a random number x, respectively. Hence, the objective function can be expressed as:

$$\text{Max} \sum_{cp \in CP} \sum_{m \in M_{Pet}} \sum_{k \in N} p_k \, Pr^{Pet}_{cp,k} \, \delta_{cp,m,k} \, x^{Pet}_m$$

$$- \sum_{cp \in CFP} \sum_{k \in N} p_k \, C^{Pet+}_{cp} \, V^{Pet+}_{cp,k} - \sum_{cp \in CFP} \sum_{k \in N} p_k \, C^{Pet-}_{cp} \, V^{Pet-}_{cp,k}$$

$$-\theta_1 \sqrt{\begin{aligned} & \sum_{cp \in CFP} \sum_{m \in M_{Pet}} (x^{Pet}_m)^2 \, \text{var}(\delta_{cp,m,k}) \text{var}(Pr^{Pet}_{cp,k}) \\ & + \sum_{cp \in CFP} \sum_{m \in M_{Pet}} (x^{Pet}_m)^2 \, \text{var}(\delta_{cp,m,k}) \mu(Pr^{Pet}_{cp,k}) \\ & + \sum_{cp \in CFP} \sum_{m \in M_{Pet}} (x^{Pet}_m)^2 \, \mu(\delta_{cp,m,k}) \text{var}(Pr^{Pet}_{cp,k}) \end{aligned}} \quad (8.9)$$

$$-\theta_2 \sqrt{\sum_{cp \in CFP} \left[C^{Pet+}_{cp}\right]^2 \text{var}(V^{Pet+}_{cp,k}) + \sum_{cp \in CFP} \left[C^{Pet-}_{cp}\right]^2 \text{var}(V^{Pet-}_{cp,k})}$$

By expanding the mean and variance terms of $Pr^{Pet}_{cp,k}$, $\delta_{cp,m,k}$, $V^{Pet+}_{cp,k}$ and $V^{Pet-}_{cp,k}$, the objective function can be recast as:

$$\text{Max} \quad \sum_{cp \in CP} \sum_{m \in M_{Pet}} \sum_{k \in N} p_k \, Pr_{cp,k}^{Pet} \, \delta_{cp,m,k} \, x_m^{Pet}$$

$$- \sum_{cp \in CFP} \sum_{k \in N} p_k \, C_{cp}^{Pet+} \, V_{cp,k}^{Pet+} - \sum_{cp \in CFP} \sum_{k \in N} p_k \, C_{cp}^{Pet-} \, V_{cp,k}^{Pet-}$$

$$-\theta_1 \sqrt{\begin{array}{l} \sum_{cp \in CP} \sum_{m \in M_{Pet}} \sum_{k \in N} (x_m^{Pet})^2 \, p_k \left[\delta_{cp,m,k} - \sum_{k \in N} p_k \, \delta_{cp,m,k} \right]^2 \left[Pr_{cp,k}^{Pet} - \sum_{k \in N} p_k \, Pr_{cp,k}^{Pet} \right]^2 \\ + \sum_{cp \in CP} \sum_{m \in M_{Pet}} \sum_{k \in N} (x_m^{Pet})^2 \, p_k \left[\delta_{cp,m,k} - \sum_{k \in N} p_k \, \delta_{cp,m,k} \right]^2 \left(\sum_{k \in N} p_k \, Pr_{cp,k}^{Pet} \right)^2 \\ + \sum_{cp \in CP} \sum_{m \in M_{Pet}} \sum_{k \in N} (x_m^{Pet})^2 \, p_k \left[Pr_{cp,k}^{Pet} - \sum_{k \in N} p_k \, Pr_{cp,k}^{Pet} \right]^2 \left(\sum_{k \in N} p_k \, \delta_{cp,m,k} \right)^2 \end{array}}$$

$$-\theta_2 \sqrt{\begin{array}{l} \sum_{cp \in CP} \left[C_{cp}^{Pet+} \right]^2 \sum_{k \in N} p_k \left[V_{cp,k}^{Pet+} - \sum_{k \in N} p_k \, V_{cp,k}^{Pet+} \right]^2 \\ + \sum_{cp \in CP} \left[C_{cp}^{Pet-} \right]^2 \sum_{k \in N} p_k \left[V_{cp,k}^{Pet-} - \sum_{k \in N} p_k \, V_{cp,k}^{Pet-} \right]^2 \end{array}}$$

(8.10)

In order to understand the effect of each term on the overall objective function of the petrochemical network, different values of θ_1 and θ_2 should be evaluated, as will be shown in the illustrative case study.

8.3
Value to Information and Stochastic Solution

Since stochastic programming adds computational burden to practical problems, it is desirable to quantify the benefits of considering uncertainty. In order to address this point, there are generally two values of interest. One is the *expected value of perfect information* (EVPI) which measures the maximum amount the decision maker is willing to pay in order to get accurate information on the future. The second is the *value of stochastic solution* (VSS) which is the difference in the objective function between the solutions of the mean value problem (replacing random events with their means) and the stochastic solution (SS) (Birge, 1982).

A solution based on perfect information would yield optimal first stage decisions for each realization of the random parameter ξ. Then the expected value of these decisions, known as "wait-and-see" (WS) can be written as (Madansky, 1960):

$$WS = E_\xi [\text{Min } z(x, \xi)]$$

However, since our objective is profit maximization, the EVPI can be calculated as:

$$EVPI = WS - SS \tag{8.11}$$

The other quantity of interest is the VSS. In order to quantify it, we first need to solve the mean value problem, also referred to as the expected value (EV) problem. This can be defined as Min $z(x, E[\xi])$ where $E[\xi] = \bar{\xi}$ (Birge, 1982). The solution of the EV problem provides the first stage decisions variables evaluated at expectation of the random realizations. The expectation of the EV problem, evaluated at different realization of the random parameters, is then defined as (Birge, 1982):

$$EEV = E_{\xi}[\text{Min } z(\bar{x}(\bar{\xi}), \xi)]$$

where $\bar{x}(\bar{\xi})$ is evaluated from the EV model, allowing the optimization problem to choose second stage variables with respect to ξ. Similarly, since our objective is profit maximization, the value of the stochastic solution can be expressed as:

$$VSS = SS - EEV \tag{8.12}$$

The value of the stochastic solution can also be evaluated as the cost of ignoring uncertainty in the problem. These concepts will be evaluated in our case study.

8.4
Illustrative Case Study

A number of case studies have been developed to demonstrate the performance of the optimization models and illustrate the effect of process yield, raw material and product prices, and lower product market demand variations. The case study in this chapter is based on Al-Sharrah, Hankinson and Elkamel (2006) and is the same as the example presented in Chapter 4. The petrochemical network included 81 processes connecting the production and consumption of 65 chemicals which gave rise to 5265 uncertain process yield parameters. In addition, the model included 11 uncertain product demand parameters and 65 uncertain parameters representing raw materials and product prices. This gives a total of 5341 uncertain parameters which were modeled with a total number of 200 scenarios for each random parameter. Due to the high number of uncertain parameters and the fact that this type of data is generally stored in spreadsheets, all scenarios were generated in Excel spreadsheets using a pseudo-random number generator. The input data was then imported to GAMS using the GAMS-Excel interface. All uncertain parameters were assumed to have a normal distribution. However, this assumption places no restriction on the generality of the proposed modeling approach as other sampling distributions can be easily used instead.

The modeling system GAMS (Brooke et al., 1996) is used to set up the optimization models. The computational tests were carried out on a Pentium M processor 2.13 GHz. The models were solved with DICOPT (Viswanathan and Grossmann, 1990). The NLP subproblems were solved with CONOPT2 (Drud, 1994), while the MILP master problems were solved with CPLEX (CPLEX Optimization Inc, 1993).

8.4.1
Solution of Stochastic Model

The two-stage mixed-integer stochastic program with recourse that includes a total number of 200 scenarios for each random parameter is considered in this section. All random parameters were assumed to follow a normal distribution and the scenarios for all random parameters were generated simultaneously. Therefore, the recourse variables account for the deviation from a given scenario as opposed to the deviation from a particular random number realization.

Table 8.1 shows the stochastic model solution for the petrochemical system. The solution indicated the selection of 22 processes with a slightly different configuration and production capacities from the deterministic case, Table 4.2 in Chapter 4. For example, acetic acid was produced by direct oxidation of n-butylenes instead of the air oxidation of acetaldehyde. Furthermore, ethylene was produced by pyrolysis of ethane instead of steam cracking of ethane–propane (50–50 wt%). These changes, as well as the different production capacities obtained, illustrate the effect of the uncertainty in process yield, raw material and product prices, and lower product

Table 8.1 Stochastic model solution.

Process selected	Production Capacity (10^3 t/year)
acetaldehyde by the one-step oxidation from ethylene	991.0
acetic acid by direct oxidation of n-butylenes	397.6
acetone by oxidation of propylene	169.6
acetylene by submerged flame process	179.7
acrylic fibers by batch suspension polymerization	245.8
acrylonitrile by cyanation/oxidation of ethylene	300.9
ABS by suspension/emulsion polymerization	419.6
benzene by hydrodealkylation of toluene	767.4
butadiene by extractive distillation	104.9
chlorobenzene by oxychlorination of benzene	146.0
cumene by the reaction of benzene and propylene	144.3
ethylbenzene by the alkylation of benzene	692.8
ethylene by pyrolysis of ethane	1051.8
hydrogen cyanide by the ammoxidation of methane	180.6
phenol by dehydrochlorination of chlorobenzene	122.7
polystyrene (crystal grade) by bulk polymerization	133.4
polystyrene (expandable beads) by suspension polymerization	102.8
polystyrene (impact grade) by suspension polymerization	154.1
poly(vinyl chloride) by bulk polymerization	407.6
styrene from ethylbenzene by hydroperoxide process	607.7
vinyl acetate from reaction of ethylene and acetic acid	113.8
vinyl chloride by the hydrochlorination of acetylene	417.8

demands. In fact, ignoring uncertainty of the key parameters in decision problems can yield non-optimal and infeasible decisions (Birge, 1995). The annual profit of the petrochemical network studied under uncertainty was found to be $2 698 552.

However, in order to properly evaluate the added-value of including uncertainty of the problem parameters, we will investigate both the EVPI and the VSS.

In order to evaluate the VSS we first solved the deterministic problem, as illustrated in the previous section, and fixed the petrochemical network and the production rate of the processes. We then solved the EEV problem by allowing the optimization problem to choose second stage variables with respect to the realization of the uncertain parameters ξ. From (8.12), the VSS is:

$$VSS = 2\,698\,552 - EEV$$
$$EEV = 2\,184\,930$$
$$VSS = 513\,622$$

This indicates that the benefit of incorporating uncertainty in the different model parameters for the petrochemical network investment is $513 622. On the other hand, the EVPI can be evaluated by first finding the "wait-and-see" (WS) solution. The latter can be obtained by taking the expectation for the optimal first stage decisions evaluated at each realization ξ. From (8.11), the EVPI is:

$$EVPI = WS - 2\,698\,552$$
$$WS = 2\,724\,040$$
$$EVPI = 25\,488$$

This implies that if it were possible to know the future realization of the demand, prices and yield perfectly, the profit would have been $2 724 040 instead of $2 698 552, yielding savings of $25 488. However, since acquiring perfect information is not viable, we will merely consider the value of the stochastic solution as the best result. These results show that the stochastic model provided an excellent solution as the objective function value was not too far from the result obtained by the WS solution.

However, as mentioned in the previous section, the stochastic model takes a decision based on first-stage and expected second-stage costs, and, hence, does not account for the decision maker risk behavior (risk-averse or risk taker). For this reason, a more realistic approach would consider higher moments where the trade-off between the mean value and the variations of different scenarios is appropriately reflected.

8.4.2
Solution of the Robust Model

Considering risk in terms of variations in both projected benefits and recourse variables provided a more robust analysis of the problem. As explained earlier, the problem will have a more robust solution as the results will remain close to optimal for all given scenarios through minimizing the variations of the projected benefit. On the other hand, the model will be more robust as minimizing the variations in the recourse variables leads to a model that is almost feasible for all the scenarios

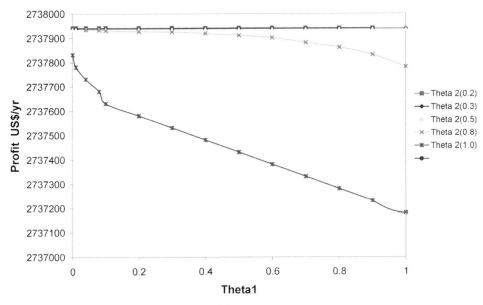

Figure 8.1 Profit variation for different values of θ_1 and θ_2.

considered. In order to investigate the effect of each term on the original problem, the spectrum of results generated by varying the scaling factors must be explored. For this reason, the model was repeatedly solved for different values of θ_1 (profit variations) and θ_2 (recourse variables variations) in order to construct the efficient frontier plot of expected profit versus risk measured by standard deviation. Figure 8.1 illustrates the change in profit with different values of profit variations, denoted by θ_1, and recourse variables variations, denoted by θ_2. The graph shows the decline in expected profit as we penalize the variations in process yield, profit, and demand by increasing the values of θ_1 and θ_2. These values will depend on the policy adopted by the investor, whether being risk-averse or risk taker, and can be read directly from the efficient frontier plots. Figure 8.2 demonstrates the trade-off between profit with risk, represented by the total standard deviation in prices and demand, with respect to different values of θ_1 and θ_2. Furthermore, it was found that the problem is more sensitive to variations in product prices than to variations in product demand and process yields for values of θ_1 and θ_2 that maintain the final petrochemical structure. As the values of θ_1 and θ_2 increase some processes became too risky to include in the petrochemical network, and, instead, importing some final chemicals became a more attractive alternative. This type of analysis requires accurate pricing structure of the local market under study as compared to the global market. In this study, however, we restricted the range of the variations of the scaling parameters θ_1 and θ_2 to values that will maintain all processes obtained from the stochastic model. This approach was adopted as the objective of the study was to include all required processes that will meet a given demand.

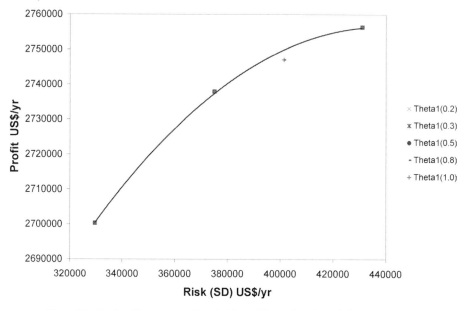

Figure 8.2 Trade-off between profit and risk at different θ_2 values while varying θ_1.

8.5
Conclusion

A robust mixed-integer nonlinear programming model for maximizing profit in the design of petrochemical networks was presented. Uncertainty in process yield, raw material and product prices, and lower product market demand were considered. In addition, operational risk was accounted for in terms of variance in projected benefits, process yield and forecasted demand. Including these different sources of uncertainty in the problem as well as modeling risk provided a more robust analysis for this type of highly strategic planning application in the chemical industry. The proposed approach increased solution robustness and model robustness by incorporating penalty terms for deviation from both projected benefits and recourse variables, respectively.

The results of the model considered in this Chapter under uncertainty and with risk consideration, as one can intuitively anticipate, yielded different petrochemical network configurations and plant capacities when compared to the deterministic model results. The concepts of EVPI and VSS were introduced and numerically illustrated. The stochastic model provided good results as the objective function value was not too far from the results obtained using the wait-and-see approach. Furthermore, the results in this Chapter showed that the final petrochemical network was more sensitive to variations in product prices than to variation in market demand and process yields when the values of θ_1 and θ_2 were selected to maintain the final petrochemical structure.

References

Al-Sharrah, G.K., Hankinson, G., and Elkamel, A. (2006) Decision-making for petrochemical planning using multiobjective and strategic tools. *Chemical Engineering Research and Design*, **84** (A11), 1019.

Birge, J.R. (1982) Stochastic linear programs with fixed recourse. *Mathematical Programming*, **24**, 314.

Birge, J.R. (1995) Models and model value in stochastic programming. *Annals of Operations Research*, **59**, 1.

Brooke, A., Kendrick, D., Meeraus, A., and Raman, R. (1996) *GAMS–A User's Guide*, GAMS Development Corporation, Washington DC.

CPLEX Optimization Inc . (1993) *Using the CPLEX Callable Library and CPLEX Mixed Integer Library*, CPLEX Optimization Inc., Incline Village, NV.

Drud, A.S. (1994) CONOPT: a large scale GRG code. *ORSA Journal of Computing*, **6**, 207.

Johnson, N.L. and Tetley, H. (1955) *Statistics*, vol. I, Cambridge University Press.

Madansky, A. (1960) Inequalities for stochastic linear programming problems. *Management Science*, **6**, 197.

Markowitz, H.M. (1952) Portfolio selection. *Journal of Finance*, **7**, 77.

Mulvey, J.M., Vanderbei, R.J., and Zenios, S.A. (1995) Robust optimization of large-scale systems. *Operations Research*, **43**, 264.

Sahinidis, N.V. (2004) Optimization under uncertainty: state-of-the-art and opportunities. *Computers & Chemical Engineering*, **28**, 971.

Viswanathan, J. and Grossmann, I.E. (1990) A combined penalty function and outer-approximation method for MINLP optimization. *Computers & Chemical Engineering*, **14**, 769.

9
Stochastic Multisite Refinery and Petrochemical Network Integration

In this final chapter, we study the multisite refinery and petrochemical integration problem, explained in Chapter 5, under uncertainty. The randomness considered includes both the objective function and right-hand side parameters of inequality constraints. As pointed out in the previous chapters, considering such strategic planning decisions requires proper handling of uncertainties as they play a major role in the final decision making process.

9.1
Introduction

The main focus of this chapter is to develop a mathematical programming tool for simultaneous design of an integrated network of refineries and petrochemical processes under uncertainty. The proposed model not only addresses the integration between the multiple refineries and devises their detailed plans, but also establishes the design of an optimal petrochemical network from a range of process technologies to satisfy a given demand. In this study we treat parameter uncertainty in terms of imported crude oil price, refinery product price, petrochemical product price, refinery market demand, and petrochemical lower level product demand. The problem is modeled as an MILP two-stage stochastic model with recourse. Furthermore, we apply the sample average approximation (SAA) method within an iterative scheme to generate the required scenarios. The solution quality is then statistically evaluated by measuring the optimality gap of the final solution. The objective function is a minimization of the annualized cost over a given time horizon among the refineries by improving the coordination and utilization of excess capacities in each facility and maximization of the added value in the petrochemical system. The proposed formulation is applied to an integrated industrial scale case study of a petrochemical complex for the production of polyvinyl chloride (PVC) and a network of petroleum refineries.

The remainder of this Chapter 9 is organized as follows. In Section 9.2 we will explain the proposed model formulation for the refinery and petrochemical integration problem under uncertainty. Then, in Section 9.3, we will explain the scenario generation methodology adopted. In Section 9.4, we present the computational results and the performance of the proposed approach on an industrial scale case study. The chapter ends with concluding remarks in Section 9.5.

9.2
Model Formulation

The proposed formulation addresses the problem of planning an integrated network of refineries and petrochemical processes. The proposed model is based on the formulation proposed in the previous chapters of this book. The general problem under study was defined in Chapter 5. In this study, uncertainty was accounted for by using two-stage stochastic programming with a recourse approach. Parameter uncertainties considered in this study included uncertainties in the imported crude oil price $CrCost_{cr}$, refinery product price Pr^{Ref}_{cfr}, petrochemical product price Pr^{Pet}_{cp}, refinery market demand $D_{Ref\ cfr}$, and petrochemical lower level product demand $D^{L}_{Pet\ cp}$. Uncertainty is modeled through the use of mutually exclusive scenarios of the model parameters with a finite number N of outcomes. For each $\xi_k = (CrCost_{cr,k}, Pr^{Ref}_{cfr,k}, Pr^{Pet}_{cp,k}, D_{Ref\ cfr,k}, D^{L}_{Pet\ cp,k})$ where $k = 1, 2, \ldots, N$, there corresponds a probability p_k. The generation of the scenarios will be explained in a later section. The proposed stochastic model is as follows:

$$\text{Min} \sum_{cr \in CR} \sum_{i \in I} \sum_{k \in N} p_k\, CrCost_{cr,k}\, S^{Ref}_{cr,i} + \sum_{p \in P} OpCost_p \sum_{cr \in CR} \sum_{i \in I} z_{cr,p,i}$$

$$+ \sum_{cir \in CIR} \sum_{i \in I} \sum_{i' \in I} InCost_{i,i'}\, y^{Ref}_{pipe\,cir,i,i'} + \sum_{i \in I} \sum_{m \in M_{Ref}} \sum_{s \in S} InCost_{m,s}\, y^{Ref}_{exp\,m,i,s}$$

$$- \sum_{cfr \in PEX} \sum_{i \in I} \sum_{k \in N} p_k\, Pr^{Ref}_{cfr,k}\, e^{Ref}_{cfr,i} - \sum_{cp \in CP} \sum_{m \in M_{Pet}} \sum_{k \in N} p_k\, Pr^{Pet}_{cp,k}\, \delta_{cp,m}\, x^{Pet}_m$$

$$+ \sum_{cfr \in CFR} \sum_{k \in N} p_k\, C^{Ref+}_{cfr}\, V^{Ref+}_{cfr,k} + \sum_{cfr \in CFR} \sum_{k \in N} p_k\, C^{Ref-}_{cfr}\, V^{Ref-}_{cfr,k}$$

$$+ \sum_{cp \in CFP} \sum_{k \in N} p_k\, C^{Pet+}_{cp}\, V^{Pet+}_{cp,k} + \sum_{cp \in CFP} \sum_{k \in N} p_k\, C^{Pet-}_{cp}\, V^{Pet-}_{cp,k}$$

where $i \neq i'$

(9.1)

Subject to

$$z_{cr,p,i} = S^{Ref}_{cr,i} \quad \forall\, cr \in CR, i \in I \quad \text{where} \quad p \in P' = \{\text{Set of CDU processes } \forall \text{ plant } i\}$$

(9.2)

$$\sum_{p \in P} \alpha_{cr,cir,i,p}\, z_{cr,p,i} + \sum_{i' \in I} \sum_{p \in P} \xi_{cr,cir,i',p,i}\, xi^{Ref}_{cr,cir,i',p,i} - Fi^{Pet}_{cr,cir \in RPI,i}$$

$$- \sum_{i' \in I} \sum_{p \in P} \xi_{cr,cir,i,p,i'}\, xi^{Ref}_{cr,cir,i,p,i'} - \sum_{cfr \in CFR} w_{cr,cir,cfr,i} - \sum_{rf \in FUEL} w_{cr,cir,rf,i} = 0 \quad (9.3)$$

$$\forall\, cr \in CR, cir \in CIR, i' \,\&\, i \in I \quad \text{where} \quad i \neq i'$$

$$\sum_{cr \in CR} \sum_{cir \in CB} w_{cr,cir,cfr,i} - \sum_{cr \in CR} \sum_{rf \in FUEL} w_{cr,cfr,rf,i} - \sum_{cr \in CR} Ff^{Pet}_{cr,cfr \in RPF,i} = x^{Ref}_{cfr,i}$$
$$\forall \, cfr \in CFR, i \in I \tag{9.4}$$

$$\sum_{cr \in CR} \sum_{cir \in CB} \frac{w_{cr,cir,cfr,i}}{sg_{cr,cir}} = xv^{Ref}_{cfr,i} \qquad \forall \, cfr \in CFR, i \in I \tag{9.5}$$

$$\sum_{cir \in FUEL} cv_{rf,cir,i} \, w_{cr,cir,rf,i} + \sum_{cfr \in FUEL} w_{cr,cfr,rf,i} - \sum_{p \in P} \beta_{cr,rf,i,p} \, z_{cr,p,i} = 0$$
$$\forall \, cr \in CR, rf \in FUEL, i \in I \tag{9.6}$$

$$\sum_{cr \in CR} \sum_{cir \in CB} \left(\begin{array}{c} att_{cr,cir,q \in Qv} \dfrac{w_{cr,cir,cfr,i}}{sg_{cr,cir}} + att_{cr,cir,q \in Qw} \\[6pt] \left[w_{cr,cir,cfr,i} - \displaystyle\sum_{rf \in FUEL} w_{cr,cfr,rf,i} - \displaystyle\sum_{cr \in CR} Ff^{Pet}_{cr,cfr \in RPF,i} \right] \end{array} \right)$$
$$\geq q^{L}_{cfr,q \in Qv} \, xv^{Ref}_{cfr,i} + q^{L}_{cfr,q \in Qw} \, x^{Ref}_{cfr,i}$$
$$\forall \, cfr \in CFR, q = \{Qw, Qv\}, i \in I \tag{9.7}$$

$$\sum_{cr \in CR} \sum_{cir \in CB} \left(\begin{array}{c} att_{cr,cir,q \in Qv} \dfrac{w_{cr,cir,cfr,i}}{sg_{cr,cir}} + att_{cr,cir,q \in Qw} \\[6pt] \left[w_{cr,cir,cfr,i} - \displaystyle\sum_{rf \in FUEL} w_{cr,cfr,rf,i} - \displaystyle\sum_{cr \in CR} Ff^{Pet}_{cr,cfr \in RPF,i} \right] \end{array} \right)$$
$$\leq q^{U}_{cfr,q \in Qv} \, xv^{Ref}_{cfr,i} + q^{U}_{cfr,q \in Qw} \, x^{Ref}_{cfr,i}$$
$$\forall \, cfr \in CFR, q = \{Qw, Qv\}, i \in I \tag{9.8}$$

$$\text{Min } C_{m,i} \leq \sum_{p \in P} \gamma_{m,p} \sum_{cr \in CR} z_{cr,p,i} \leq \text{Max } C_{m,i} + \sum_{s \in S} AddC_{m,i,s} \, y^{Ref}_{\exp \, m,i,s}$$
$$\forall \, m \in M_{Ref}, i \in I \tag{9.9}$$

$$\sum_{cr \in CR} \sum_{p \in P} \xi_{cr,cir,i,p,i'} \, xi^{Ref}_{cr,cir,i,p,i'} \leq F^{U}_{cir,i,i'} \, y^{Ref}_{pipe \, cir,i,i'}$$
$$\forall \, cir \in CIR, i' \, \& \, i \in I \quad \text{where} \quad i \neq i' \tag{9.10}$$

$$\sum_{i \in I} \left(x^{Ref}_{cfr,i} - e^{Ref}_{cfr',i} \right) + V^{Ref+}_{cfr,k} - V^{Ref-}_{cfr,k} = D_{Ref \, cfr,k} \qquad \forall \, cfr \in CFR \quad cfr' \in PEX \quad k \in N$$
$$\tag{9.11}$$

$$IM_{cr}^L \le \sum_{i\in I} S_{cr,i}^{Ref} \le IM_{cr}^U \quad \forall\, cr \in CR \tag{9.12}$$

$$Fn_{cp\in NRF}^{Pet} + \sum_{i\in I}\sum_{cr\in CR} Fi_{cr,cp\in RPI,i}^{Pet} + \sum_{i\in I}\sum_{cr\in CR} Ff_{cr,cp\in RPF,i}^{Pet}$$
$$+ \sum_{m\in M_{Pet}} \delta_{cp,m}\, x_m^{Pet} + V_{cp\in CFP,k}^{Pet+} - V_{cp\in CFP,k}^{Pet-} = D_{Pet\; cp\in CFP,k}^L + xi_{cp\in CIP}^{Pet}$$
$$\forall\, cp \in CP \quad k \in N \tag{9.13}$$

$$Fn_{cp\in NRF}^{Pet} + \sum_{i\in I}\sum_{cr\in CR} Fi_{cr,cp\in RPI,i}^{Pet} + \sum_{i\in I}\sum_{cr\in CR} Ff_{cr,cp\in RPF,i}^{Pet}$$
$$+ \sum_{m\in M_{Pet}} \delta_{cp,m}\, x_m^{Pet} \le D_{Pet\; cp\in CFP}^U \quad \forall\, cp \in CP \tag{9.14}$$

$$B_m^L\, y_{proc\, m}^{Pet} \le x_m^{Pet} \le K^U\, y_{proc\, m}^{Pet} \quad \forall\, m \in M_{Pet} \tag{9.15}$$

$$\sum_{cp\in CIP} y_{proc\, m}^{Pet} \le 1 \quad \forall\, m \in M_{Pet}\; \text{that produces}\; cp \in CIP \tag{9.16}$$

$$\sum_{cp\in CFP} y_{proc\, m}^{Pet} \le 1 \quad \forall\, m \in M_{Pet}\; \text{that produces}\; cp \in CFP \tag{9.17}$$

$$Fn_{cp}^{Pet} \le S_{cp}^{Pet} \quad \forall\, cp \in NRF \tag{9.18}$$

The above formulation is an extension of the deterministic model explained in Chapter 5. We will mainly explain the stochastic part of the above formulation. The above formulation is a two-stage stochastic mixed-integer linear programming (MILP) model. Objective function (9.1) minimizes the first stage variables and the penalized second stage variables. The production over the target demand is penalized as an additional inventory cost per ton of refinery and petrochemical products. Similarly, shortfall in a certain product demand is assumed to be satisfied at the product spot market price. The recourse variables $V_{cfr,k}^{Ref+}$, $V_{cfr,k}^{Ref-}$, $V_{cp,k}^{Pet+}$ and $V_{cp,k}^{Pet-}$ in Equations 9.11 and 9.13 represent the refinery production shortfall and surplus as well as the petrochemical production shortfall and surplus, respectively, for each random realization $k \in N$. These variables will compensate for the violations in Equations 9.11 and 9.13 and will be penalized in the objective function using appropriate shortfall and surplus costs C_{cfr}^{Ref+} and C_{cfr}^{Ref-} for the refinery products, and C_{cp}^{Pet+} and C_{cp}^{Pet-} for the petrochemical products, respectively. Uncertain parameters are assumed to follow a normal distribution for each outcome of the random realization ξ_k. Although this might sound restrictive, this assumption imposes no limitation on the generality of the proposed approach as other distributions can be easily incorporated instead. Furthermore, in Equation 9.13 an additional term xi_{cp}^{Pet} was added to the left-hand-side representing the flow of intermediate petrochemical stream of $cp \in CIP$. This term may be set to zero under the assumption that intermediate petrochemical streams produced by any process are consumed within the petrochemical network. However, this assumption may not be valid when considering a subsystem of the petrochemical network.

9.3
Scenario Generation

The solution of stochastic problems is generally very challenging as it involves numerical integration over the random continuous probability space of the second stage variables (Goyal and Ierapetritou, 2007). An alternative approach is the discretization of the random space using a finite number of scenarios. A common approach is the Monte Carlo sampling where independent pseudo-random samples are generated and assigned equal probabilities (Ruszczyński and Shapiro, 2003). The sample average approximation (SAA) method, also known as stochastic counterpart, is employed. The SAA problem can be written as (Verweij et al., 2003):

$$v_N = \min_{x \in X} c^T x + \frac{1}{N} \sum_{k \in N} Q(x, \xi^k) \qquad (9.19)$$

It approximates the expectation of the stochastic formulation (usually called the "true" problem) and can be solved using deterministic algorithms. Problem (9.19) can be solved iteratively in order to provide statistical bounds on the optimality gap of the objective function value. The iterative SAA procedure steps are explained in Section 7.5 of Chapter 7.

9.4
Illustrative Case Study

This section presents the computational results of the proposed model and sampling scheme. We examine the same case study considered in Chapter 5 of the three refineries and the PVC complex. We consider uncertainty in the imported crude oil price, refinery product price, petrochemical product price, refinery market demand, and petrochemical lower level product demand. The major capacity constraints for the refinery network are given in Table 9.1 and the process technologies considered for the production of PVC are listed in Table 9.2. The representation for the topology of the refineries network and petrochemical technologies for the PVC production are given in Figures 5.2 and 5.3, respectively, in Chapter 5. In the presentation of the results, we focus on demonstrating the sample average approximation computational results as we vary the sample sizes and compare their solution accuracy and the CPU time required for solving the models.

The modeling system GAMS (Brooke et al., 1996) is used for setting up the optimization models. The computational tests were carried out on a Pentium M

Table 9.1 Major refinery network capacity constraints.

Production capacity	Higher limit (10^3 t/year)		
	R1	R2	R3
Distillation	45 000.	12 000.0	9900.0
Reforming	700.0	2000.0	1800.0
Isomerization	200.0	—	450.0
Fluid catalytic cracker	800.0	1400.0	—
Hydrocracker	—	1800.0	2400.0
Delayed coker	—	—	1800
Des gas oil	1300.0	3000.0	2400.0
Des cycle gas oil	200.0	750.0	—
Des ATK	—	1200.0	1680.0
Des distillates	—	—	450.0
Crude availability			
Arabian Light		31 200.0	
Local demand			
LPG		$N(432,20)$	
LN		—	
PG98		$N(400,20)$	
PG95		$N(4390,50)$	
JP4		$N(2240,50)$	
GO6		$N(4920,50)$	
ATK		$N(1700,50)$	
HFO		$N(200,20)$	
Diesel		$N(400,20)$	
Coke		$N(300,20)$	

processor 2.13 GHz and the MILP problems were solved with CPLEX (CPLEX Optimization Inc, 1993).

The problem was solved for different sample sizes N and N' to illustrate the variation of optimality gap confidence intervals, while fixing the number of replications R to 30. The replication number R need not be very large to get an insight into $\bar{v}_N$ variability. Table 9.3 shows different confidence interval values of the optimality gap when the sample size of N assumes values of 1000, 2000, and 3000 while varying N' from 5000, 10 000, to 20 000 samples. The sample sizes N and N' were limited to these values due to increasing computational effort. In our case study, we ran into memory limitations when N and N' values exceeded 3000 and 20 000, respectively. The solution of the three refineries network and the PVC complex using the SAA scheme with $N = 3000$ and $N' = 20 000$ required 1114 CPU s to converge to the optimal solution.

Table 9.4 depicts the results of the optimal integration network between the three refineries and the PVC petrochemical complex. As shown in Table 9.4, the proposed

Table 9.2 Major products and process technologies in the petrochemical complex.

Product	Sale price ($/ton)	Process technology	Process index	Min Econ. Prod. (10^3 t/year)
Ethylene (E)	$\mathcal{N}$ (1570,10)	Pyrolysis of naphtha (low severity)	1	250
		Pyrolysis of gas oil (low severity)	2	250
		Steam cracking of naphtha (high severity)	3	250
		Steam cracking of gas oil (high severity)	4	250
Ethylene dichloride (EDC)	$\mathcal{N}$ (378,10)	Chlorination of ethylene	5	180
		Oxychlorination of ethylene	6	180
Vinyl chloride monomer (VCM)	$\mathcal{N}$ (1230,10)	Chlorination and Oxychlorination of ethylene	7	250
		Dehydrochlorination of ethylene dichloride	8	125
Polyvinyl chloride (PVC)	$\mathcal{N}$ (1600,10)	Bulk polymerization	9	50
		Suspension polymerization	10	90

Table 9.3 Computational results with SAA for the stochastic model.

		Lower bound sample size = N		
		1000	2000	3000
UB sample size $N' = 5000$	Number of samples: $R = 30$			
	LB estimate: $\bar{v}_N$	8 802 837	8 804 092	8 804 456
	LB error: $\tilde{\varepsilon}_l(\alpha = 0.975)$	3420	2423	1813
	UB estimate: $\hat{v}_{N'}$	8 805 915	8 805 279	8 805 578
	UB error: $\tilde{\varepsilon}_u(\alpha = 0.975)$	7776	7715	7778
	95% Conf. interval	[0,14274]	[0,11324]	[0,10713]
	CPU (s)	65	112	146
$N' = 10 000$	LB estimate: $\bar{v}_N$	8 800 071	8 802 080	8 804 305
	LB error: $\tilde{\varepsilon}_l(\alpha = 0.975)$	3356	2527	2010
	UB estimate: $\hat{v}_{N'}$	8 803 310	8 803 204	8 803 414
	UB error: $\tilde{\varepsilon}_u(\alpha = 0.975)$	5473	5833	5410
	95% Conf. interval	[0,12068]	[0,9484]	[0,7420]
	CPU (sec)	196	224	263
$N' = 20 000$	LB estimate: $\bar{v}_N$	8 796 058	8 801 812	8 802 511
	LB error: $\tilde{\varepsilon}_l(\alpha = 0.975)$	3092	2345	1755
	UB estimate: $\hat{v}_{N'}$	8 802 099	8 804 121	8 802 032
	UB error: $\tilde{\varepsilon}_u(\alpha = 0.975)$	3837	3886	3880
	95% Conf. interval	[0,12970]	[0,8540]	[0,5635]
	CPU (s)	1058	1070	1114

Table 9.4 Stochastic model results of refinery and petrochemical networks.

	Process variables		Results (10^3 t/year)		
			R1	R2	R3
Refinery	Crude oil supply		4500.0	12 000.0	9900.0
	Production levels	Crude unit	4500	12 000	9900
		Reformer	612.5	1824.6	1784.6
		Isomerization	160	—	450
		FCC	378	1174.2	—
		Hydrocracker	—	1740.4	2400
		Delayed coker	—	—	1440
		Des gas oil	1300	3000	2400
		Des cycle gas oil	168.6	600	—
		Des ATK	—	1200	1654.8
		Des distillates	—	—	366.2
	Intermediate streams exchange	From R1 VGO	—	—	576.1 to HCU
		R2 LN	—	—	112.4 to Isom
		R3 VGO	—	274.8 to FCC	—
	Exports	PG95	439.8		
		JP4	1101.9		
		GO6	2044.2		
		HFO	1907.8		
		ATK	1887.6		
		Coke	110.7		
		Diesel	5.1		
Petrochemical	Refinery feed to PVC complex	Gas oil	788.6	1037.0	71.3
	Production levels	S. Crack GO (4)	486.8		
		Cl & OxyCl E (7)	475.4		
		Bulk polym. (9)	220.0		
	Final products	PVC	220.0		
Total cost ($/yr)			$8 802 000		

model redesigned the refinery integration network topology and operating policies when compared to the deterministic solution obtained in Chapter 5. However, similar to the deterministic solution the model selected gas oil, an intermediate refinery stream, as the refinery feedstock to the petrochemical complex as opposed to the typically used light naphtha feedstock. This selection emphasizes the importance of sparing the light naphtha stream for the gasoline pool to get maximum gasoline production.

PVC production, on the other hand, is carried out by first high severity steam cracking of gas oil to produce ethylene. Vinyl chloride monomer (VCM) is then produced through the chlorination and oxychlorination of ethylene and finally, VCM is converted to PVC by bulk polymerization. The annual production cost across the refineries and the PVC complex was $8 802 000.

9.5
Conclusion

In this chapter, a two-stage stochastic mixed-integer programming model for designing an integration and coordination policy among multiple refineries and a petrochemical network under uncertainty was described. Uncertainty was considered in the parameters of imported crude oil price, refinery product price, petrochemical product price, refinery market demand, and petrochemical lower level product demand. The approach employs the sample average approximation method with a statistical bounding and validation technique. In this sampling scheme, a relatively small sample size N is used to make decisions, with multiple replications, and another independent larger sample is used to reassess the objective function value while fixing the first stage variables. The proposed model performance was illustrated on a network of three large-scale refineries and a PVC petrochemical complex. The formulation captured the simultaneous design of both the refinery and petrochemical networks and illustrated the economic potential and trade-offs. The consideration of uncertainties in this type of high level strategic planning model, especially with the current volatile market environment, presented an adequate treatment of the problem and a proper optimization tool.

References

Brooke, A., Kendrick, D., Meeraus, A., and Raman, R. (1996) *GAMS–A User's Guide*, GAMS Development Corporation, Washington DC.

CPLEX Optimization Inc (1993) *Using the CPLEX Callable Library and CPLEX Mixed Integer Library*, CPLEX Optimization Inc., Incline Village, NV.

Goyal, V. and Ierapetritou, M.G. (2007) Stochastic MINLP optimization using simplicial approximation. *Computers and Chemical Engineering*, **31** 1081.

Ruszczyński, R. and Shapiro, A. (2003) *Handbooks in Operations Research and Management Science-Volume 10: Stochastic Programming*, Elsevier Science, Amsterdam, Netherlands.

Verweij, B., Ahmed, S., Kleywegt, A.J., Nemhauser, G., and Shapiro, A. (2003) The sample average approximation method applied to stochastic routing problems: A computational study. *Computational Optimization & Applications*, **24**, 289.

Appendix A:
Two-Stage Stochastic Programming

In a standard two-stage stochastic programming model, decision variables are divided into two groups; namely, first stage and second stage variables. First stage variables are decided upon before the actual realization of the random parameters. Once the uncertain events have unfolded, further design or operational adjustments can be made through values of the second-stage or alternatively called recourse variables at a particular cost. This concept of recourse has had many applications to linear, integer, and non-linear programming.

A standard formulation of the two-stage stochastic linear program is:

$$\text{Min}_x c^T x + E[Q(x, \xi(\omega))]$$
$$\text{subject to} \quad Ax = b, x \geq 0 \tag{A.1}$$

where $Q(x, \xi(\omega))$ is the optimal value of the second stage problem:

$$\text{Min}_x q^T y$$
$$\text{subject to} \quad Tx + Wy = h, y \geq 0 \tag{A.2}$$

where x and y are vectors of the first and second stage decision variables, respectively. The second stage problem depends on the data $\xi = (q, h, T, W)$ where any or all elements can be random. The expectation in (A.1) is with respect to the probability distribution of $\xi(\omega)$. Matrices T and W are called technological and recourse matrices, respectively. The second stage problem (A.2) can be considered as penalty for the violation of the constraint $Tx = h$.

There are two different ways of representing uncertainty. The first approach is the continuous probability distribution where numerical integration is employed over the random continuous probability space. This approach maintains the model size but on the other hand introduces nonlinearities and computational difficulties to the problem. The other approach is the scenario-based approach where the random space is considered as discrete events. The main disadvantage of this approach is the substantial increase in computational requirements with an increase in the number of uncertain parameters. The discrete distribution with a finite number K of possible outcomes (scenarios) $\xi_k = (q_k, h_k, T_k, W_k)$ corresponds to the probability p_k. Hence, Equations (A.1) and (A.2) can be written as a deterministic equivalent problem and represented as follows:

$$\text{Min}_{x,y_1,\ldots,y_k} c^T x + \sum_{k=1}^{K} p_k q_k^T y_k$$

subject to $\quad Ax = b$ \hfill (A.3)

$$T_k x + W_k y_k = h_k \quad k = 1, \ldots, K$$

$$x \geq 0, y_k \geq 0 \quad k = 1, \ldots, K$$

Due to the complexity of numerical integration and the exponential increase in sample size with the increase of the random variables, we employ an approximation scheme know as the sample average approximation (SAA) method, also known as stochastic counterpart. The SAA problem can be written as:

$$v_N = \min_{x \in X} c^T x + \frac{1}{N} \sum_{k \in N} Q(x, \xi^k) \quad (A.4)$$

It approximates the expectation of the stochastic formulation (usually called the "true" problem) and can be solved using deterministic algorithms.

Appendix B:
Chance Constrained Programming

The philosophy of the previous methods of stochastic programming was to ensure feasibility of the problem through the second-stage problem at a certain penalty cost. In the chance-constrained approach, some of the problem constraints are expressed probabilistically, requiring their satisfaction with a probability greater than a desired level. This approach is particularly useful when the cost and benefits of second-stage decisions are difficult to assess as the use of second-stage or recourse actions is avoided. These intangible components include loss of goodwill, cost of off-specification products and outsourcing of production.

For a typical linear programming model:

$$\text{Min}_x c^T x \text{ subject to } Ax \geq b, \ x \geq 0 \tag{B.1}$$

assume that there are uncertainties in the matrix A (left-hand-side coefficient) and in the right-hand-side vector b and the above constraint must be satisfied with a probability $p \in (0,1)$. Then the probabilistic model can be expressed as follows:

$$\text{Min}_x c^T x \quad \text{subject to} \quad P(Ax \geq b) \geq p, \ x \geq 0 \tag{B.2}$$

If we consider a single constraint, for the sake of simplicity, then the above becomes $P(a^t x \geq b) \geq p$. Furthermore, assume the randomness is only in the right-hand-side with a distribution of F. When $F(\beta) = p$, then the constraint can be written as $F(a^t x) \geq p \rightarrow a^t x \geq \beta$. In this case, the model yields a standard linear program.

Appendix C:
SAA Optimal Solution Bounding

Here we present a proof of the SAA bound on the true optimal solution of the stochastic problem. The proof is rather intuitive for the upper bound since it starts from a feasible solution. However, the lower bound proof is more involved and is as follows:

Proposition
For any R independent and identically distributed sample batches (denoting the number of sample replication) each with sample size of N, that is, $\xi^{j1}, \ldots, \xi^{jN}, j = 1, \ldots, R$, the $E[v_N^j] \leq v^*$ is always valid.

Proof
For any feasible point x' that belongs to the solution set X, the inequality below is valid:

$$\hat{f}_N(x') \geq \min_{x \in X} \hat{f}_N(x)$$

By taking the expectations of both sides and minimizing the left-hand-side, we get:

$$\min_{x \in X} E[\hat{f}_N(x)] \geq E[\min_{x \in X} \hat{f}_N(x)]$$

Since $E[\hat{f}_N(x)] = f(x)$, it follows that $v^* \geq E[v_N^j]$.

Index

a

accuracy, solutions 148
added-value maximization 84
algorithms
– back-propagation 37
– generalized Benders decomposition 59
– successive disaggregation 140
alteration 9–10
alternative refinery uses 14
analysis methodology 121–123
aniline point 30
annualized cost 94–97
approximation
– non-linear objective functions 82
– programming 7
– sample average, see sample average approximation
aromatics 12–13
artificial neural networks (ANN) 36–44
atmospheric bottom 6, 8
availability constraints 113
average approximation, sample, see sample average approximation
average boiling point, volumetric 30

b

back-propagation algorithm 37
balances, material, see material balances
Benders decomposition algorithm 59
benzene 12, 93
binary variables 93, 98
blending 5
– product 10
block flow diagram, refinery 4
– refinery 45, 113
BMCI (Bureau of Mines Correlation Index) 13
boiling point, volumetric average 30
bottom, atmospheric 6, 8
bounding
– optimal 187
– statistical 139
BTX (benzene, toluene, xylene) 12–13, 15–16, 93
Bureau of Mines Correlation Index (BMCI) 13

c

calculation, variance 116–117
capacity
– limitation and expansion 64–65
– plant 45
capacity constraints
– multisite refinery planning 155
– refinery network 178
– single refinery planning 150
carbon Conradson residue (CCR) 28, 40
catalytic cracking, fluid, see fluid catalytic cracking
catalytic processes 9–10
CDU (crude distillation unit) 45, 52
chance constrained programming 185
characterization factor, Watson 28, 43
chemical plant, block flow diagram 113
chloride
– polyvinyl 91, 99, 102–103, 173, 177–180
– VCM 104, 180–181
coking processes 8–9
common purpose model 68
complexity, computational 58
computational time 151
conditional value-at-risk 144
confidence interval 151
configuration, refinery 5, 7
Conradson, carbon Conradson residue (CCR) 28, 40

continuous refinery plants   3
coordination, multisite refinery network   55–79
correlation, FCC processes   27–31
correlation index, BMCI   13
cost
– annualized   94–97
– fixed-charge   61
– recourse penalty   128–130
counterpart, stochastic, *see* sample average approximation
CPLEX   48, 103
cracking processes   9, 25–31
– ethylene   92
– steam   102
– stream   15
cross plots   38
crude distillation unit (CDU)   45, 52
crude oil distillation   6
crude supply   5
cyclo-paraffins   13–14

d

decision-maker, risk-neutral   144
decision variables, first/second stage   183
decomposition algorithm, Benders   59
dedicated processes   56
degree of uncertainty   127
delayed coking   8
demand constraints   113
demand uncertainty   117–118
deterministic model
– planning   19–137
– variation coefficient   123
deviation, mean-absolute   120, 133, 144
disaggregation algorithm, successive   140
distillation   7–8
– crude oil   6
distillation unit, crude   45, 52
distribution, normal   151, 163, 166–167, 176
distribution function, joint probability   119
downstream processes   86
dual price   50–51

e

economic and environmental analysis   83
economic risk factor   132
efficient frontier plot   135, 159
efficient frontier portfolio   116
empirical models   24
engineering, process systems   111
enterprise-wide planning   22

environmental analysis   83
equipment network, state, *see* state equipment network
estimator, variance   147
ethylene cracker operations   92
ethylene feedstock shift   99
evaluation
– objective function   115
– planning models   31
excess capacities   94
exchange, intermediate streams   104–105, 157
expansion, capacity   64–65
expansion problems   57
expectation model   117–119, 125–126
expected value of perfect information (EVPI)   161, 165–166
expected value of profit   115
exterior sampling   146

f

feed API   28, 43
feedstock
– multiple   72–75
– petrochemical   12–14
– single   70–72
feedstock shift   99
final product, primary/secondary   84
financial risk   119, 132
first stage decision variables   183
fixed-charge cost   61
fixed plant yield   46
flexible processes   56
flow diagram, refinery   4, 45, 113
flow rate
– intermediate streams   64
– trans-shipment   97
fluid catalytic cracking (FCC)   9, 14–16
– ANN-based modeling   36–44
– linear programming   31–36
– process correlation   27–31
– regression-based planning   24–36
– ten-lump   26
fractionator   25
frontier, efficient   116, 135, 159
fuel gas upgrade   16
fuel system material balance   63
functions
– non-linear   120
– normal distribution, *see* normal distribution
– objective, *see* objective function
– penalty   117–118
– probability distribution   119

g

GAMS   48, 66, 84, 103
GAMS-Excel interface   166
gap, optimality   148, 152, 156
gas
– fuel   16
– natural   11
gasoline, pyrolysis   10, 15
generalized Benders decomposition algorithm   59
generic representation of risk   144

h

heating oil   47
hidden layer   37
hierarchy, process operations   22
historic operations practices   23–24
hybrid stochastic programming   112
hydrocracker   92

i

import constraint   65
increased market demand   74–75
industry, petrochemical, *see* petrochemical industry
inequality constraint   116
information, value to   165–166
initial state   67–70
input layer   38
input–output coefficient matrix   98
integration
– multisite refinery network   55–79
– network   56
– petrochemical network   91–107, 173–181
integration superstructure   62
interior sampling   146
intermediate streams   73–74
– exchange   104–105, 157
– flow rates   64
inventory requirements   113
isomerization unit   101
iterative SAA procedure   177

j

joint probability distribution function   119

l

layer, hidden   37
limitation
– capacity   64–65
– raw material   46
linear programming (LP)   5
– FCC processes   31–36
– mixed-integer, *see* mixed-integer linear programming
– stochastic model formulations   123
linear yield vectors   61
'lost work'   82

m

MAD (mean-absolute deviation)   120, 133, 144
market demand
– increased   74–75
– uncertainty   124
market requirements   46
Markowitz's mean-variance (MV) model   114, 117, 144
material balances   32, 45–46
– fuel system   63
– multisite refinery network   62–63
– stochastic programming   114
mathematical programming   5–7
mean, upper partial   141
mean-absolute deviation (MAD)   120, 133, 144
mean-risk model   119
mean-variance (MV) model, Markowitz's   114, 117, 144
mean-variance portfolio optimization   112
mechanistic models   24
method of approximation programming (MAP)   7
mixed-integer linear programming (MILP)   55–58
– multiperiod   57
– petrochemical network planning   81–89
– refinery and petrochemical industry network integration   93
– robust planning   141–144
– two-stage   161–170, 173
mixed-integer non-linear programming (MINLP)   59, 140–142, 161
mixer   67–70
models
– ANN-based   36–44
– common purpose   68
– empirical   24
– evaluation   31
– expectation, *see* expectation model
– Markowitz's MV   114, 117, 144
– mean-risk   119
– mechanistic   24
– non-linear regression   31
– planning, *see* planning models
– risk, *see* risk model
– robust   144–146, 168–170

- robustness 121–122, 142
- solutions 48
- stochastic 142–144
- two-stage stochastic 162–163, 167–168

Monte Carlo sampling 177
multiobjective optimization approach 131
multiple feedstocks 72–75
multiple replications 181
multisite refinery
- capacity constraints 155
- network integration 55–79
- network planning 69
- objective function 65–66
- petrochemical network integration 91–107
- robust planning 139–160
- stochastic 173–181

multistage stochastic optimization 140

n

natural gas 11
network
- artificial neural 36–44
- state equipment, *see* state equipment network

network integration
- petrochemical 91–107, 173–181
- processes 56

network planning
- petrochemical 81–89
- robust 139–170

non-linear function 120
non-linear objective functions 82
non-linear programming (NLP) 57, 148, 152, 166
- mixed-integer, *see* mixed-integer non-linear programming

non-linear regression models 31
normal distribution 151, 163, 166–167, 176
normal paraffins 13–14

o

objective function 46
- coefficients 49
- evaluation 115
- multisite refinery network 65–66
- non-linear 82
- refinery and petrochemical industry network integration 93–97
- robust planning 142–144
- stochastic programming 114
- two-stage MILP 163–165

oil, heating 47
olefins 13
on-line scheduling 22

one task–one equipment (OTOE) assignment 61
operational risk 119, 132
operational risk factor 128–130
operations hierarchy 22
operations practices, historic 23–24
operations research 57
optimal solution bounding 187
optimality gap 148, 152, 156
optimistic scenario 124–125
optimization
- mean-variance portfolio 112
- multiobjective 131
- robust 88, 163–165

output layer 38

p

paraffins 13–14
partial mean, upper 141
penalty cost, recourse 128–130
penalty functions 117–118
perfect information, expected value of 161, 165–166
pessimistic scenario 124–125
petrochemical feedstock 12–14
petrochemical industry 3–17
- supply chain 92
- synergy benefits 14

petrochemical network integration 91–107
- stochastic multisite refinery 173–181

petrochemical network planning 81–89
- robust 161–170

petroleum chain 94
petroleum consumption 81
petroleum refinery, planning 21–54
petroleum refining 3–17
pipelines installation 71, 94
planning
- deterministic models 19–137
- enterprise-wide 22
- model evaluation 31
- multisite refinery 155
- multisite refinery network 69
- petrochemical network 81–89
- petroleum refinery 21–54
- production 21–23
- regression-based 24–36
- robust, *see* robust planning
- single refinery 66–68, 150
- strategic 75
- model types 24–36
- yield based 44

planning under uncertainty 109–181
- single refinery plants 111–138

plant capacity 45
polyvinyl chloride (PVC) 91, 99, 102–103
– petrochemical network integration 173, 177–180
– process technologies 179
portfolio optimization, mean-variance 112
price
– dual/shadow 50–51
– random coefficients 116
– uncertainty 124
primary final product (PF) 84
primary raw material (PR) 84
probability distribution function, joint 119
process systems engineering (PSE) 111
processes
– catalytic 9–10
– coking 8–9
– correlation 27–31
– dedicated/flexible 56
– distillation 7–8
– downstream 86
– FCC 25–31
– integration 14–15
– network integration 56
– operations hierarchy 22
– PVC petrochemical complex 179
– thermal 8–9
– treatment 10
– yield uncertainty 118–119
product, primary/secondary 84
product blending 5, 10
product demand 65
product quality 63–64
product yields, uncertainty 124
production
– capacity constraints 113
– planning 21–23
– shortfall/surplus 122, 127–131
– single-site planning 22
production units 5
profit, expected value of 115
programming
– chance constrained 185
– linear, *see* linear programming
– MAP 7
– mathematical 5–7
– mixed-integer linear, *see* mixed-integer linear programming
– non-linear, *see* non-linear programming
– stochastic, *see* stochastic programming
– two-stage stochastic 183–184
pumping 94

PVC (polyvinyl chloride) 91, 99, 102–103
– petrochemical network integration 173, 177–180
– process technologies 179
pyrolysis 101
pyrolysis gasoline (pygas) 10, 15

q
quality, product 63–64

r
random price coefficients 116
raw material
  limitation 46
– primary/secondary 84
reaction network, ten-lump FCC 26
recourse penalty cost 128–130
recourse variables 163, 169
refinery
– alternative uses 14
– block flow diagram 4, 45, 113
– continuous plants 3
– multisite, *see* multisite refinery
– network capacity constraints 178
– petrochemical synergy benefits 14
– single, *see* single refinery
– standard configuration 5
– supply chain 92
refinery configuration 7
refining
– mathematical programming 5–7
– petroleum 3–17
regenerator 25
regression-based planning 24–36
regression models, non-linear 31
regulatory control 22
residue, carbon Conradson 28
risk
– financial/operational 119, 132
– generic representation 144
risk factor, operational 128–130
risk management 141
risk model 114–117, 119–121, 124–133
risk-neutral decision-maker 144
robust model 144–146
– solutions 168–170
robust optimization 88, 163–165
robust planning
– multisite refinery network 139–160
– petrochemical network 161–170
– single refinery 148–153
robustness, model/solution 121–122, 142

s

sample average approximation (SAA)   139, 146–158, 173, 177
– optimal solution bounding   187
– two-stage stochastic programming   184
sampling
– interior/exterior   146
– methodology   115
– Monte Carlo   177
scaling factors   145
scaling parameters   169
scenario
– generation   177
– optimistic/pessimistic   124–125
scheduling   21–23
– on-line   22
second stage decision variables   183
secondary final product (SF)   84
secondary raw material (SR)   84
SEN (state equipment network)   61
– PVC petrochemical complex   103
– refinery and petrochemical industry network integration   99–100
– refinery layout   67–70
– robust planning   149
sensitivity analysis   49–51
– stochastic programming   133–134
sequence control   22
shadow price   50–51
shortfall, production   122, 127–131
single feedstock   70–72
single refinery   44
– capacity constraints   150
– planning   66–68
– planning under uncertainty   111–138
– robust planning   148–153
single-site production planning   22
slack variables   51, 118
slurry settler   25
solutions
– accuracy   148
– models   48
– optimal bounding   187
– robust model   168–170
– robustness   121–122, 142
– stochastic   165–166
– trial   129
– two-stage stochastic model   167–168
– VSS   161, 165–166
– 'wait-and-see'   165–166, 168
SP, see stochastic programming
splitter/mixer   67–70
standard deviation, see variance
standard refinery configuration   5
state equipment network (SEN)   61
– PVC petrochemical complex   103
– refinery and petrochemical industry network integration   99–100
– refinery layout   67–70
– robust planning   149
statistical bounding techniques   139
steam cracking   102
stochastic counterpart, see sample average approximation
stochastic models
– multisite refinery network   142–144
– two-stage   162–163, 167–168
– variation coefficient   123
stochastic multisite refinery, petrochemical network integration   173–181
stochastic optimization, two-stage/multistage   140
stochastic programming (SP)   111
– two-stage   114, 183–184
stochastic solutions   165–166
– value of the   161
strategic planning   75
stream cracking   15
streams, intermediate   64, 73–74
successive disaggregation algorithm   140
superstructure, integration   62
supervisory control   22
supply chain   56
– refinery and petrochemical industry   92
surplus, production   122, 127–131
synergy benefits, petrochemical   14

t

tar   11
ten-lump FCC reaction network   26
thermal processes   8–9
time horizon   94
toluene   12, 93
trans-shipment flow rate   97
transportation   5
treatment processes   10
trial solutions   129
'true' problem   146–148, 177, 184
two-stage SP framework   114
two-stage stochastic model   173
– robust planning   162–163
– solutions   167–168
two-stage stochastic optimization   140
two-stage stochastic programming   183–184

u

uncertainty
– demand   117–118

– EVPI/VSS   168
– planning, *see* planning under uncertainty
– prices/market demands/product yields   124
– process yield   118–119
– PVC petrochemical complex   174
upper partial mean (UPM)   141
utilities integration   15–16

v

value-at-risk   144
value of perfect information, expected   161, 165–166
value of the stochastic solution (VSS)   161, 165–166
value to information   165–166
variables
– binary   93, 98
– first/second stage decision   183
– recourse   163, 169
– slack   51, 118
variance
– as a risk measure   141
– calculation   116–117
– estimator   147
– mean-variance portfolio optimization   112
variation coefficient   122–123
vectors, yield   61
vinyl chloride monomer (VCM)   104, 180–181
volumetric average boiling point   30
'wait-and-see' solution   165–166, 168

w

Watson characterization factor   28, 43
work, 'lost'   82

x

xylene   12, 93

y

yield
– fixed plant   46
– uncertainty   118–119, 124
yield based planning   44
yield vectors, linear   61